//n# Mysteries of Bee Explained

M. Quinby

Alpha Editions

This edition published in 2024

ISBN : 9789361472824

Design and Setting By
Alpha Editions
www.alphaedis.com
Email - info@alphaedis.com

As per information held with us this book is in Public Domain.
This book is a reproduction of an important historical work. Alpha Editions uses the best technology to reproduce historical work in the same manner it was first published to preserve its original nature. Any marks or number seen are left intentionally to preserve its true form.

Contents

PREFACE. ....- 1 -
CHAPTER I. ....- 5 -
CHAPTER II. ....- 8 -
CHAPTER III. ....- 35 -
CHAPTER IV. ....- 49 -
CHAPTER V. ....- 63 -
CHAPTER VI. ....- 71 -
CHAPTER VII. ....- 74 -
CHAPTER VIII. ....- 80 -
CHAPTER IX. ....- 87 -
CHAPTER X. ....- 93 -
CHAPTER XI. ....- 98 -
CHAPTER XII. ....- 104 -
CHAPTER XIII. ....- 108 -
CHAPTER XIV. ....- 136 -
CHAPTER XV ....- 144 -
CHAPTER XVI. ....- 150 -
CHAPTER XVII. ....- 152 -
CHAPTER XVIII. ....- 158 -
CHAPTER XIX. ....- 163 -
CHAPTER XX. ....- 173 -
CHAPTER XXI. ....- 175 -
CHAPTER XXII. ....- 186 -
CHAPTER XXIII. ....- 202 -
CHAPTER XXIV. ....- 205 -

CHAPTER XXV .... - 208 -
Footnotes .... - 217 -

PREFACE.

Before the reader decides that an apology is necessary for the introduction of another work on bees into the presence of those already before the public, it is hoped that he will have the patience to examine the contents of this.

The writer of the following pages commenced beekeeping in 1828, without any knowledge of the business to assist him, save a few directions about hiving, smoking them with sulphur, &c. Nearly all the information to be had was so mingled with erroneous whims and notions, that it required a long experience to separate essential and consistent points. It was *impossible* to procure a work that gave the information necessary for practice. From that time to the present, no sufficient guide for the inexperienced has appeared. European works, republished here, are of but little value. Weeks, Townley, Miner, and others, writers of this country, within a few years, have given us treatises, valuable to some extent, but have entirely neglected several chapters, very important and essential to the beginner. Keeping bees *has* been, and is now, by the majority, deemed a hazardous enterprise. The ravages of the moth had been so great, and loss so frequent, that but little attention was given to the subject for a long time. Mr. Weeks lost his entire stock three times in fifteen years. But soon after the discovery was promulgated, that honey could be taken from a stock without destroying the bees, an additional attention was manifest, increasing to a rage in many places. It seems to be easily understood, that *profit* must attend success, in this branch of the farmer's stock; inasmuch as the "bees work for nothing and find themselves." This interest in bees should be encouraged to continue till enough are kept to collect all the honey now wasted; which, compared with the present collections, would be more than a thousand pounds to one. But to succeed, that is the difficulty. Some eighteen years since, after a propitious season, an aged and esteemed friend said to me, "It is not to be expected that you will have such luck always; you must expect they will run out after a time. I have always noticed, when people have first-rate luck for a time, that the bees generally take a turn, and are gone in a few years."

I am not sure but, to the above remarks, may be traced the cause of my subsequent success. It stimulated me to observation and inquiry. I soon found that good seasons were the "lucky" ones, and that many lost in an adverse season, all they had before gained. Also, that strong families were the only ones on which I could depend for protection against the moth. This induced the effort to ascertain causes tending to diminish the size of families, and the application of remedies. Whether success has attended my efforts or not, the reader can judge, after a perusal of the work.

It is time that the word "*luck*," as applied to beekeeping, was discarded. The prevailing opinion, that bees will prosper for one person more than another, under the same circumstances, is fallacious. As well might it be applied to the mechanic and farmer. The careless, ignorant farmer, might occasionally succeed in raising a crop with a poor fence; but would be liable, at any time, to lose it by trespassing cattle. He might have suitable soil in the beginning, but without knowledge, for the proper application of manures, it might fail to produce; unless a *chance* application *happened* to be right.

But with the *intelligent* farmer the case is different: fences in order, manures judiciously applied, and with propitious seasons, he makes a sure thing of it. Call him "*lucky*" if you please; it is his knowledge, and care, that render him so. So with bee-keeping, the careful man is the "lucky" one. There can be no effect without a preceding cause. If you lose a stock of bees, there is a cause or causes producing it, just as certain as the failure of a crop with the unthrifty farmer, can be traced to a poor fence, or unfruitful soil. You may rest assured, that a rail is off your fence of management somewhere, or the proper applications have not been made. In relation to bees, these things may not be quite so apparent, yet nevertheless true. Why is there so much more uncertainty in apiarian science than other farming operations? It must be attributed to the fact, that among the thousands who are engaged in, and have studied agriculture, perhaps not more than one has given his energies to the nature and habits of bees. If knowledge is elicited in the same ratio, we ought to have a thousand times more light on one subject than the other, and still there are some things, even in agriculture, that may yet be learned.

It is supposed, by many, that we already have all the knowledge that the subject of *bees* affords. This is not surprising; a person that was never furnished with a full treatise, might arrive at such conclusions. Unless his own experience goes deeper, he can have no means of judging what is yet behind.

In conversation relative to this work, with a person of considerable scientific attainments, he remarked, "You do not want to give the natural history of bees at all; that is already sufficiently understood." And how is it understood; as Huber gives it, or in accordance with some of our own writers? If we take Huber as a guide, we find many points recently contradicted. If we compare authors of our day, we find them contradicting each other. One recommends a peculiarly constructed hive, as just the thing adapted to their nature and instincts. If a single point is in accordance with their nature, he labors to twist all the others to his purpose, although it may involve a fundamental principle impossible to reconcile. Some one else succeeds in another point, and proceeds to recommend something altogether different. False and contradictory assertions are made either through ignorance, or interest. Interest may blind the judgment, and spurious history may deceive.

It is folly to expect success in bee-keeping for any length of time, without a correct knowledge of their nature and instincts; and this we shall never obtain by the course hitherto pursued. As much of their labor is performed in the dark, and difficult to be observed, it has given rise to conjecture and false reasoning, leading to false conclusions.

When *I* say a thing *is so*, or say it is *not so*, what evidence has the reader that it is proved or demonstrated? *My* mere assertions are not expected to be taken in preference to another's; of such proof, we have more than enough. Most people have not the time, patience, or ability, to set down quietly with close observation, and investigate the subject thoroughly. Hence it has been found easier to receive error for truth, than to make the exertion necessary to confute it; the more so, because there is no guide to direct the investigation. I shall, therefore, pursue a different course; and for every *assertion* endeavor to give a test, that the reader may apply and satisfy himself, and trust to no one. As for theories, I shall try to keep them separate from facts, and offer such evidence as I have, either for or against them. If the reader has further proof that presents the matter in another light, of course he will exercise the right to a difference of opinion.

I could give a set of rules for practice, and be very brief, but this would be unsatisfactory. When we are told a thing *must be done*, most of us, like the "inquisitive Yankee," have a desire to know *why* it is necessary; and then like to know *how* to do it. This gives us confidence that we are right. Hence, I shall endeavor to give the practical part, in as close connection with the natural history, that dictates it, as possible.

This work will contain several chapters entirely new to the public: the result of my own experience, that will be of the utmost value to all who desire to realize the greatest possible advantages from their bees.

The additions to chapters already partially discussed by others, will contain much original matter not to be found elsewhere. When many stocks are kept, the chapter on "Loss of Queens," alone, will, with attention, save to any one, not in the secret, enough in one season to be worth more in value than many times the cost of this work. The same might be said of those on diseased brood, artificial swarms, wintering bees, and many others.

If such a work could have been placed in my hands twenty years ago, I should have realized hundreds of dollars by the information. But instead of this, my course has been, first to suffer a loss, and then find out the remedy, or preventive; from which the reader may be exempt, as I can confidently recommend these directions.

Another new feature will be found in the duties of each season being kept by itself, commencing with the spring and ending with the winter management.

In my anxiety to be understood by all classes of readers, I am aware that I have made the elegant construction and arrangement of sentences of secondary importance; therefore justly liable to criticism. But to the reader, whose object is information on this subject, it can be of but little consequence.

CHAPTER I.

A BRIEF HISTORY.

THREE KINDS OF BEES.

Every prosperous swarm, or family of bees, must contain one queen, several thousand workers, and, part of the year, a few hundred drones.

QUEEN.   WORKER.   DRONE.

QUEEN DESCRIBED.

The queen is the mother of the entire family; her duty appears to be only to deposit eggs in the cells. Her abdomen has its full size very abruptly where it joins the trunk or body, and then gradually tapers to a point. She is longer than either the drones or workers, but her size, in other respects, is a medium between the two. In shape she resembles the worker more than the drone; and, like the worker, has a sting, but will not use it for anything below royalty. She is nearly destitute of down, or hairs; a very little may be seen about her head and trunk. This gives her a dark, shining appearance, on the upper side—some are nearly black. Her legs are somewhat longer than those of a worker; the two posterior ones, and the under surface, are often of a bright copper color. In some of them a yellow stripe nearly encircles the abdomen at the joints, and meets on the back. Her wings are about the same as the workers, but as her abdomen is much longer, they only reach about two-thirds the length of it. For the first few days after leaving the cell, her size is much less than after she has assumed her maternal duties. She seldom, perhaps never, leaves the hive, except when leading a swarm, and when but a few days old, to meet the drones, in the air, for the purpose of fecundation. The manner of the queen's impregnation is yet a disputed point, and probably never witnessed by any one. The majority of close observers, I believe, are of opinion that the drones are the males, and that sexual connection takes place in the air,[1] performing their amours while on the wing, like the humble-bee and some other insects. It appears that one impregnation is operative during her life, as old queens are not afterwards seen coming out for that purpose.

DESCRIPTION AND DUTY OF WORKERS.

As all labor devolves on the workers, they are provided with a sack, or bag, for honey. Basket-like cavities are on their legs, where they pack the pollen of flowers into little pellets, convenient to bring home. They are also provided with a sting, and a virulent poison, although they will not use it abroad when unmolested, but, if attacked, will generally defend themselves sufficient to escape. They range the fields for honey and pollen, secrete wax, construct combs, prepare food, nurse the young, bring water for the use of the community, obtain propolis to seal up all crevices about the hive, stand guard, and keep out intruders, robbers, &c., &c.

DESCRIPTION OF DRONES.

When the family is large and honey abundant, a brood of drones is reared; the number, probably, depends on the yield of honey, and size of the swarm, more than anything else. As honey becomes scarce, they are destroyed. Their bodies are large and rather clumsy, covered with short hairs or bristles. Their abdomen terminates very abruptly, without the symmetry of the queen or worker. Their buzzing, when on the wing, is louder, and altogether different from the others. They seem to be of the least value of any in the hive. Perhaps not more than one in a thousand is ever called upon to perform the duty for which they were designed. Yet they assist, on some occasions, to keep up the animal heat necessary in the old hive after a swarm has left.

MOST BROOD IN SPRING.

In spring and first of summer, when nearly all the combs are empty, and food abundant, they rear brood more extensively than at any other period, (towards fall more combs are filled with honey, giving less room for brood.) The hive soon becomes crowded with bees, and royal cells are constructed, in which the queen deposits her eggs. When some of these young queens are advanced sufficiently to be sealed over, the old one, and the greater part of her subjects, leave for a new location, (termed swarming.) They soon collect in a cluster, and, if put into an empty hive, commence anew their labors; constructing combs, rearing brood, and storing honey, to be abandoned on the succeeding year for another tenement. One in a hundred may do it the same season, if the hive is filled and crowded again in time to warrant it. Only large early swarms do this.

THEIR INDUSTRY.

Industry belongs to their nature. When the flowers yield honey, and the weather is fine, they need no impulse from man to perform their part. When their tenement is supplied with all things necessary to reach another spring, or their store-house full, and no necessity or room for an addition, and we supply them with more space, they assiduously toil to fill it up. Rather than to waste time in idleness, during a bounteous yield of honey, they have been known to deposit their surplus in combs outside the hive, or under the stand. This natural industrious habit lies at the foundation of all the advantages in bee-keeping; consequently our hives must be constructed with this end in view; and at the same time not interfere with other points of their nature; but this subject will be discussed in the next chapter. Those peculiar traits in their nature, mentioned in this, will be more fully discussed in different parts of this work, as they appear to be called for, and where proof will be offered to sustain the positions here assumed, which as yet are nothing more than mere assertions.

CHAPTER II.

HIVES.

HIVES TO BE THOROUGHLY MADE.

Hives should be constructed of good materials, boards of good thickness, free from flaws and cracks, well fitted and thoroughly nailed.

The time of making them is not very particular, providing it is done in season. It certainly should not be put off till the swarming period, to be made as wanted, because if they are to be painted; it should be done as long as possible before, as the rank smell of oil and paint, just applied, might be offensive to the bees.

But what kind of hive shall be made?

In answer, some less than a thousand forms have been given. The advantages of bee-keeping depend as much upon the construction of hives, as any one thing; yet there is no subject pertaining to them on which there is such a variety of opinions, and I have but little hopes of reconciling all these conflicting views, opinions, prejudices, and interests.

DIFFERENT OPINIONS ABOUT THEM.

One is in favor of the old box, and the cruel practice of killing the bees to obtain the honey, as the only means to obtain "luck;" "they are sure to run out if they meddle with them." Another will rush to the opposite extreme, and advocate all the extravagant fancies of the itinerant patent-vender, as the *ne plus ultra* of all hives, when perhaps it would be worth more for fire-wood than the apiary.

THE AUTHOR HAS NO PATENT TO RECOMMEND.

To remove from the mind of the reader all apprehension that I am about condemning one patent to recommend another, I would say in the beginning, that I have *no patent to praise, no interest in deceiving*, and I hope no prejudices to influence me, in advocating or condemning *any* system. I wish to make bee-keeping plain, simple, economical, and profitable; so that when we sum up the profit "it shall not be found in the other pocket."

It is a principle recognized by our statute, that no person is suitable as a Juror, who is biased either by interest or prejudice. Now whether I am the impartial Jurist, is not for me to say: but I wish to discuss the subject fairly. I hope some few will be enabled to see their own interest: at any rate, dismiss prejudice, as far as possible, while we examine wherein *one class* in community is unprofitable to bee-keepers.

SPECULATORS SUPPORTED LONG ENOUGH.

We have faithfully supported a host of speculators on our business for a long time; often not caring one straw about our success, after pocketing the fee of successful "humbuggery." One is no sooner gone, than we are beset by another, with something altogether different, and of course the acme of perfection.

PREFIX OF PATENT A BAD RECOMMENDATION.

This has been done until the very prefix of patent, or premium, attached to a hive, renders it almost certain that there must be something deleterious to the apiarian; either in expense of construction or intricate and perplexing in management, requiring an engineer to manage, and a skilful architect to construct.

What does the American savage, who without difficulty can track the panther or wolf, know of the principles of chemistry? What does the Chemist know of following a track in the forest, when nothing but withered leaves can guide him? Each understands principles, the *minutiæ* of which the other never dreamed.

IGNORANCE OF OFFICERS AND COMMITTEES.

Thus it appears to be with granting patents and premiums, if we take what has been patented and praised by our committees and officers as improvements in bee-culture. These men may be capable, intelligent, and well fitted for their sphere, but in bee matters, about as capable of judging, as the Hottentot would be of the merits of an intricate steam-engine. Knowledge and experience are the only qualifications competent to decide.

OPPOSITION TO SIMPLICITY.

I am aware that among the thousands whose direct interest is opposed to my simple, plain manner of getting along, many will be ready to contend with me for every departure from their patent, improved or premium hives, as the case may be.

BY GAINING ONE POINT, PRODUCE ANOTHER EVIL.

I think it will be an easy matter to show that every departure from simplicity to gain *one* point, is attended in another by a correspondent evil, that often exceeds the advantage gained. That we have made vast improvements in art and science, and in every department of human affairs, no one will deny; consequently, it is assumed we must correspondingly improve in a bee-hive; forgetting that nature has fixed limits to the instinct of the bee, beyond which she will not go!

It will be necessary to point out the advantages and objections to these pretended improvements, and then we will see if we cannot avoid the objections, *and retain the advantages, without the expense*, by a simple addition to the common hive; because if we expect to encourage bee-keeping, they must have better success than a neighbor of mine, who expended fifty dollars for bees and a patent, and lost all in three years! Most bee-keepers are farmers; very few are engineers sufficient to work them successfully. I would say to all such as do not understand the nature of bees, adhere to simplicity until you do, and then I am quite sure you will have no desire for a change.

FIRST DELUSION.

Probably the first delusion in the patent line originated with the idea, that to obtain surplus honey, it was absolutely necessary to have a chamber hive. To get rid of the depredations of mice, the suspended hive was contrived. The inclined bottom-board was then added to throw out the worms. To prevent the combs from sliding down, the lower end was contracted.

The principle of bees rearing queens from worker-eggs when destitute, gave rise to the dividing hive in several forms. Comb, when used several years, becomes thickened and black, and needs changing; hence the changeable hives, Non-swarmers have been introduced to save risk and trouble. Moth-proof hives to prevent the ravages of worms, &c., &c.

CHAMBER HIVE.

The chamber hive is made with two apartments; the lower and largest is for the permanent residence of the bees, the upper or chamber for the boxes. Its merits are these: the chamber affords all the protection necessary for glass boxes; considered as a cover, it is never lost. Its demerits are inconvenience in handling; it occupies more room if put in the house in the winter; if glass boxes are used, only one end can be seen, and this may be full when the other may hold some pounds yet, and we cannot possibly know until it is taken out. I know we are told to return such boxes when not full "and the bees will soon finish them," but this will depend on the yield of honey at the time; if abundant, it will be filled; if not, they will be very likely to take a hint, and remove below what there is in the box; whereas if the chamber was separate from the hive, and was not a chamber but a loose cap to cover the boxes, it could be raised at any time without disturbing a single bee, and the precise time of the boxes being filled ascertained, (that is, when they are of glass.)

MRS. GRIFFITH'S HIVE.

Mrs. Griffith, of New Jersey, is said to have invented the suspended chamber hive with the inclined bottom-board. One would suppose this was sufficiently inconvenient to use, and difficult and expensive to construct.

WEEKS' IMPROVEMENT.

Yet Mr. Weeks makes an alteration, calls it an improvement, the expense is but a trifle more; it is sufficient to be sanctioned by a patent. From front to rear, the bottom is about three inches narrower than the top, somewhat wedge-shape; it has the merit to prevent the combs from slipping down, when they *happen* to be made, to have the edges supported. The objections are, that filth from the bees will not fall as readily to the bottom as if every side was perpendicular, and the extra trouble in constructing.

INCLINED BOTTOM-BOARDS DO NOT THROW OUT ALL THE WORMS.

Inclined bottom-boards form the basis of one or two patents, said to be good to roll out the worms. I can imagine a pea rolling off such a board; but a worm is not often found in a rolling condition. Most of us know, that when a worm drops from the combs, it is like the spider, with a thread attached above. The only way that I can imagine one to be thrown out by these boards, is to have it dead when it strikes it, or so cold that it cannot spin a thread, and wind to shake the board, till it rolls off. The objections to these boards are coupled with the suspended hive, with which they are usually connected.

OBJECTIONS TO SUSPENDED HIVES.

All suspended hives *must be objectionable* to any one who wishes to know the *true* condition of his bees at all times. Only think of the trouble of unhooking the bottom-board, and getting down on your back, or twisting your neck till your head is dizzy, to look up among the combs, and then see nothing satisfactory for want of light; or to lift the hive from its supporters, and turn it over. The operation is too formidable for an indolent man, or one that has much other business. The examination would very probably be put off till quite sure it would do no longer, and sometimes a few days after that, when you will very often find your bees past remedy.

SEE BEES OFTEN.

"*See your bees often*," is a choice recipe,—it is worth five hundred dollars at interest, even when you have but few stocks. How necessary then that we have every facility for a close and minute inspection. How much easier to turn up a hive that simply rests on a stand. Sometimes it is necessary to turn the hive, even bottom up, and let the rays of the sun directly among the combs, to see *all* the particulars. By this close inspection, I have often ascertained the cause of some difficulty, and provided a remedy, thus saving a good many that in a short time would have been lost; yet, with a little help, were as valuable as any by another year.

HALL'S PATENT.

Mr. Hall has added a lower section to his hive, about four inches deep, with two boards inside, like the roof of a house, to discharge the worms, &c.; but as these boards would interfere with close inspection, they are objectionable. Several other variations of inclined bottom-boards and suspended hives have been contrived, to obtain a patent, but the objections offered will apply to most of them. I shall not weary the reader by noticing in detail *every* hive that has been patented; I think if I notice the *principles of each kind*, it will test his patience sufficiently.

JONES' PATENT.

Jones' dividing hive was probably suggested by this instinctive principle of the bee, viz.: when a stock by any accident loses its queen, and the combs contain eggs or very young larvæ, they will rear another. Now if a hive is constructed so as to divide the brood-combs, it would seem quite certain that the half without a queen, would raise one; and we could multiply our stocks without swarms, the trouble of hiving, and risk of their going to the woods, &c.

AN EXPERIMENT.

Several years ago, I thought I had obtained a principle that would revolutionize the whole system of bee management. In 1840 I constructed such hives, and put in the bees to test by actual experiment, the utility of what seemed so very plausible in theory. It would appear that this principle suggested the same idea to Mr. Jones; perhaps with this difference: I think he did not wait to test the plan thoroughly, before obtaining his patent in '42. One vender of rights asserted that 63 stocks were made from one in three years; but somehow a great many that obtained the rights, failed in their expectations. From my experiments, I think I could guess at some of the reasons.

Mr. A.—"Well, what are the reasons? give us your experience, if you please, I am interested; I had the right for such a hive, and had a lot made to order, that cost more money in the end than I shall ever pay again for anything about bees."

Do not be too hasty, friend, I think I can instruct you to keep bees on principles in accordance with their nature, which is very simple, so that if you can be induced to try again, we will have the *hives* cost but little, at any rate.

REASONS OF FAILURE IN DIVIDING HIVES.

The greatest difficulty with dividing hives, appeared to be here. It must be constructed with a partition or division to keep the combs in each apartment separate; otherwise, we make tearing work in the division. When bees are first put into such hives, unless the swarm is very large, and honey abundant, one apartment will be filled to the bottom before a commencement is made in the other.

Mr. A.—"What difference can that make? It is necessary to have the hive full; if it cannot be all filled at once, why let them fill part."

The difference is this. The first combs built by a swarm are for brood, and store-combs afterwards, as needed; one apartment will be nearly filled with all brood-combs, and the other with store-combs and honey. Now in the two kinds of cells there is a great difference; those for breeding are near half an inch in length, while those for storing are sometimes two inches or more; totally unfit for breeding; until the bees cut them off to the proper length, which they will not do, unless compelled for want of room, consequently this side of store-combs is but little used for brood. When such hive is divided, the chances are not more than one in four, that this apartment will have any young bees of the proper age from which to raise a queen; if not, and the old queen is in the part with the brood-comb, where she will be ninety-nine times in a hundred, one half of the hive is lost for want of a queen.

Mr. A.—"Ah! I think I now understand how I lost one-half of nearly every hive I divided. I also lost some of them in the winter; there was plenty of bees as well as honey; can you tell the cause of this?"

I will guess that they starved.

Mr. A.—"Starved! why, I said there was plenty of honey."

I understood it, but nevertheless feel quite sure.

Mr. A.—"I would like to see that made plain; I can't understand how they could starve when there was honey!"

CAUSE OF STARVING IN SUCH HIVES.

I said one apartment would be filled with brood-combs; this will be occupied, at least partially, with brood as long as the yield of honey lasts; consequently, there will be but little room for storing here, but the other side may be full throughout. The bees will take up their winter quarters among the brood-combs. Now suppose the honey in this apartment is all exhausted during a severe turn of cold weather, what can the bees do? If one should leave the mass and go among the frosty combs for a supply, its fate would be as certain as starvation. Without frequent intervals of warm weather to melt all frost on the combs, and allow the bees to go into the other apartment for honey, they *must* starve.

The cost of construction is another objection to this hive, as the labor bestowed on one is more than would finish two, that would be much better.

ADVANTAGES OF THE CHANGEABLE HIVE CONSIDERED.

The value of changeable hives is based upon the following principle:—Each young bee when it first hatches from the egg, is neither more nor less than a worm; when it receives the necessary food, the bees seal it over; it will then spin a cocoon, or line its cell with a coating of silk, less in thickness than the thinnest paper: this remains after the bee leaves it. It is evident, therefore, that after a few hundreds have been reared in a cell, and each one has left its cocoon, that such cell must be somewhat diminished, although the thickness of a dozen cocoons could not be measured; and this old cell needs removing, that the bees may replace it with a new one. But how shall it be done? This is a feat for the display of ingenuity. A common man might go about it in a very sensible, simple manner, might possibly turn the hive over, and cut out the old combs when necessary, without knowing perhaps that the patent-vender could *sell* a receipt to do the thing *scientifically*, the benefit of which would be many times on the principle of a surgeon cutting off your head, to get a good chance to tie a small artery according to system; or would show you a roundabout way of half a dozen miles to accomplish what the same number of rods would do. Had we not ocular demonstration of the fact, we could not suppose so many variations for the same end could be invented. But if we reward ingenuity, it will be stimulated to great exertions. Perhaps if we describe the merits of one or two of this class, the utility of this principle may be comprehended.

VARIATION OF THESE HIVES.

First, then, the sectional hive of various patterns has been patented; it consists generally of about three boxes, one above another; the top of each has one large hole, or several small ones, or cross-bars, about an inch wide, and half an inch apart; these holes or spaces allowing the bees to pass from one box to the other. When all are full, the upper one is removed, and an empty one put under the bottom; in this way all are changed, and the combs renewed in three years; very easily and quietly done. This is as far as a patent-vender wishes the subject investigated; and some of his customers have not gone beyond this point. As an offset for these advantages, we will first look at the cost of such hive.

EXPENSE IN CONSTRUCTING CHANGEABLE HIVES.

It is as much work to construct each separate section, as a common hive; consequently, it is three times the expense to begin with. It is objectionable for wintering bees, on the same principle as the dividing hive. I object to it on another point: our surplus honey will never be pure, as each section must be used for breeding, and every cell so used, will contain cocoons corresponding to the number of bees raised.

SURPLUS HONEY WILL CONTAIN BEE-BREAD.

Also pollen, or bee-bread, is always stored in the vicinity of the young brood; some of this will remain mixed with the honey, to please the palate with its *exquisite flavor*. The majority will probably prefer all surplus honey stored in pure comb, where it will be with proper management.

I will here give a full description of a hive on this principle, as I have the description from one of its advocates, in the Dollar Newspaper, Philadelphia: called Cutting's Patent Changeable Hive.

DESCRIPTION OF CUTTING'S CHANGEABLE HIVE.

"The size of the changeable hive most used in this section, has an outside shell, made of inch boards, about two feet high and sixteen and a half inches square, with a door hung in the rear. On the inside are three boxes or drawers, which will hold about one thousand cubic inches each, and when filled with honey, usually weigh about thirty-five pounds, which is a sufficient amount of honey to winter a large swarm. The sides of these drawers are made of boards, about half an inch thick; the tops and bottoms of the lower drawers and ends of the upper drawers should be three-fourths of an inch, and the drawers should be fourteen inches high, fourteen inches from front to rear, and six and three-fourths inches wide. Two of these drawers stand side by side, with the third placed flatwise upon the two, with a free communication from one drawer to another, by means of thirty three-fourth inch holes on the side of each drawer, and twenty-four in the bottom of the upper drawer, and holes in the top and bottom of the lower drawers, to correspond, and slides to cut off the communication when occasion may require. Thus we see our hive may be one hive, with communication sufficiently free throughout, or we may have three hives combined. The drawers have tubes made in them, (for the bees to pass and repass), which are made to go through the front side of the hive. The back-side of the drawers are doors, with glass set in them. These drawers set up from the bottom of the hive, and rest on pieces of wood, closely fitted in such a way, as to make a space under the drawers for the *dirt, dead bees*, and *water*, which collect in the bottom of hives in winter; between the drawers and the outside is an air space of about one-third of an inch.

These hives, when well made and painted, will last many years, and those doing much in the business will find it an advantage to have a few extra drawers. Having given you some idea of the construction of the changeable hive, I will proceed to notice some of the most important reasons why I prefer this hive to any I have yet seen. First because the hive, being constructed upon the changeable principle, so that by taking out a full drawer, and placing an empty one in its stead, our comb is always kept new, wherefore, the size of the bee is preserved, and kept in a more healthy, or prosperous state, or condition, than when obliged to remain and continue to breed, in the old comb, when the cells have become small. Secondly, because small, late swarms may be easily united. Thirdly, because large swarms may be easily divided. Fourthly, because however late a swarm may come off, it may be easily supplied with honey for the winter, by taking from a full hive a surplus drawer, and placing it in the hive of the late swarm. Fifthly, because a column of air between the drawers and the outside of the hive is a non-conductor of both heat and cold, preventing the melting of the comb, and securing the bees against frost and cold."

Now here is a full description of perhaps as good a hive as any of its class; it is given for the benefit of those who wish to go miles instead of rods; they may know the road, especially as they can have the privilege by paying for it: for myself, I had rather be excused,—why, reading the description has nearly exhausted my patience; what should I do if I attempted to make one?

FIRST OBJECTION, COST OF CONSTRUCTION.

The first obstacle in the way (after the right is obtained) is the construction. Let's see; we want inch boards to make the shell, three-quarter inch boards for the tops and bottoms of drawers, half inch for sides, hinges to hang a door, glass for back of drawers, tubes for the egress of the bees, and slides to cut off communication. It will be necessary to get a mechanic, and a workman too. Those 108 holes that must be bored, *must match*, or it is of no use to make them. But few farmers would have the tools requisite, a still less number the skill and patience to do it. What the cost might be by the time a hive was ready to receive the bees, I could not say; but guess it might be some three or four dollars.

HIVES CAN BE MADE WITH LESS EXPENSE.

The one I shall recommend, without paint, will not cost, or need not, over 37-1/2 cents, with cover, etc. Now, if we wish hives for ornament, it is well enough to expend something for the purpose; but it is well not to refine too much, as there are limits which, if passed, will render it unfit for bees. Therefore, when profit is an object, the extra expense will or ought to be made up by the bees, in return for an expensive domicil. But will they do it? The merits of the one under consideration are fully given. "First, by taking out a full drawer and putting in an empty one in its stead, the combs are always kept new, and cells of full size." Now this fear of bees becoming dwarfs in consequence of being reared in cells too small, has done more mischief among the bees, and their owners' pockets, than if the fact had never been thought, or heard of.

OLD BREEDING CELLS WILL LAST A LONG TIME.

These old cells do not need renewing half as often as has been represented. It is the interest of these patent-venders to sell rights; this interest either blinds their eyes as to facts, or lulls the internal monitor of right, while acquisitiveness is gratified. The same cells can be used for breeding six or eight years, perhaps longer, and no one can tell the difference by the size of the bees; I have two stocks now in their tenth year without renewal of comb. A neighbor of mine kept a stock twelve years in the same combs; it proved as prosperous as any. I have heard of their lasting twenty, and am inclined to believe it.

CELLS LARGER THAN NECESSARY AT FIRST.

The bees seem to make a provision for this emergency, the sheets of comb are farther apart than actually necessary at first, the diameter of the cell is also a little larger than the size of the young bee requires. *Of this we are certain*—great many young bees *can* be raised in a cell, and not be diminished in size, sufficient to be detected. The bottom fills up faster than at the sides, and as they do so, the bees add a little to the length, until the ends of these cells on two parallel combs approximate too close to allow the bees to pass freely; before which time it is unnecessary to remove comb for being old.

EXPENSE OF RENEWING COMBS.

One important item should be considered in this matter, by those who are so eager for new combs. It is doubtful whether one in 500 ever thought of the expense of renewing comb. I find it estimated by one writer,[2] that twenty-five lbs. of honey was consumed in elaborating about half lb. wax. This without doubt is an over estimate, but no one will deny that some is used.

BEST TO USE OLD COMBS AS LONG AS THEY WILL ANSWER.

I am satisfied of this much, from actual experience, that every time the bees have to renew their brood-combs in a hive, they would make from ten to twenty-five lbs. in boxes, hence I infer that their time can be more profitably employed than in constructing brood-combs *every year*. I would also suggest that when combs have been once used for breeding it is the best use they can be applied to, after that, as the cocoons render it unfit for much else than a little wax.

METHOD FOR PRUNING WHEN NECESSARY.

But when the combs do actually need removing, I prefer the following method of pruning, to driving the bees out entirely, as has been recommended. It can be done in about an hour. As we are comparing the merits of different methods of getting rid of old combs, I shall give mine here, notwithstanding it may seem a little out of place.

The best time is a little before night. The first movement is to blow under the hive some tobacco smoke (the best means of charming them I ever found); the bees, deprived of all disposition to sting, retreat up among the combs to get away from the smoke; now raise the hive from the stand and carefully turn it bottom upwards, avoiding any jar, as some of the bees that were in the top when the smoke was introduced, and did not get a taste, will now come to the bottom to ascertain the cause of the disturbance; these should receive a share, and they will immediately return to the top, perfectly satisfied. When so many bees are in the hive, as to be in the way in pruning, (which if there is not it is not worth it,) get an empty hive the size of the old one, and set it over, stopping the holes; now strike the lower hive with a hammer or stick, lightly and rapidly, five or ten minutes, when nearly all the bees will be in the upper hive, and set that on the stand. There being now nothing in the way, except a few scattering bees, that I will *warrant not to sting, unless you pinch or get them fast.*

TOOLS FOR CUTTING OUT COMB.

The broad one is very readily made from a piece of an old scythe, about 18 inches long, by any blacksmith, by simply taking off the back, and forming a shank for a handle at the heel. The end should be ground all on one side, and square across like a carpenter's chisel. This is for cutting down the sides of the hive; the level will keep it close the whole length, when you wish to remove all the combs; it being square instead of pointed or rounded, no difficulty will be found in guiding it,—it being very thin; no combs are mashed by crowding.

The other tool is for cutting off combs at the top or any other place. It is merely a rod of steel three-eighths of an inch diameter, about two feet long, with a thin blade at right angles, one and a half inches long, and a quarter inch wide, both edges sharp, upper side bevelled, bottom flat, &c. You will find these tools very convenient; be sure and get them by all means, the cost cannot be compared to the advantages.

Now with the tools just described, proceed to remove the brood-combs from the centre of the hive. The combs near the top and outside are used but little for breeding, and are generally filled with honey; these should be left as a good start for refilling, but take out all that is necessary, while you are about it; then reverse the hives, putting the one containing the bees under the other; by the next morning all are up; now put it on the stand, and this job is done without one cent extra expense for a patent to help you, and the bees are much better off for the honey left, which has to be taken away with all patent plans that I have seen, and this, as has been remarked, is not worth much, occupied as it is with a few cocoons and bee-bread. It is worth much more to the bees, and they will give us pure comb and honey for it.

USE OF TOBACCO SMOKE.

"I would not do it for fifty dollars, the bees would sting me to death." Stop a moment, if you never tried the efficacy of tobacco smoke, you know nothing of a powerful agent; this is the grand secret of success; without it, I admit it would be somewhat hazardous; but with it, I have done it time after time without receiving a single sting, and no protection whatever, for either hands or face.

But is there no difficulty with our sectional or changeable hive, when this feat is to be performed? The combs will be made in the two drawers similar to the dividing hive, brood-combs in one side, and store-combs in the other. We wish to remove the one with brood-combs of course, (as that is the one where the combs are thick and bad, &c.) Where will the queen be? With the brood-comb, where her duty is most likely to be; well, this is the one we want, and we take it out. How is she to get back? She must go back, or we have three chances in four of losing the stock; but her majesty will remain perfectly easy, as well as some of the workers, wherever you put the drawer.

FURTHER OBJECTIONS TO A SECTIONAL HIVE.

I can see no other way but to break the box, look her up, and help the helpless thing home, (the chances of being stung may be here too.) Now, for a time at least, they must use the other drawer for breeding, where most of the cells are unfit. There is altogether too great a proportion of drone-cells; these, as well as the other size, will nearly all be much too long, and will have to be cut off to the proper length, a waste of wax as well as labor. Another thing might be set down per disadvantage of Mr. Cutting's hive; the job of getting a swarm into such hive, at first, I fancy would not be desirable to many. Now, when we strike the balance, putting expense, difficulties, and perplexities on one side, and simplicity and economy on the other, it appears like a "great cry for little wool." But stop a moment, four other advantages are enumerated in its favor: second, third, and fourth are borrowed from the common hive, or are all available here when required. But fifthly, allows a "column of air between the drawers and outside of the hive, is a non-conductor of heat and cold," &c. This is an advantage not possessed by the common hive; neither does the common hive offer such advantages to the moth, by affording such snug quarters for worms to spin their cocoons, when they cannot be destroyed without considerable trouble.

NON-SWARMERS.

Here I will endeavor to be brief; I feel anxious to get through with this disagreeable part, where every word I say will clash with somebody's interest or prejudice. The merits of this hive are to obtain surplus honey with but little trouble, which often succeeds in satisfying people of its utility. The principal objection is found on the score of profit. Suppose we start with one, call it worth five dollars in the beginning, at the end of ten years it is worth no more, very likely not as much, (the chances of its failing, short of that time, we will not take into the account;) we might get annually, say five dollars worth of surplus honey, amounting to fifty dollars.

CONTRAST OF PROFIT.

The swarming hive, we suppose, will throw off one swarm annually, and make us one dollar's worth of surplus honey, (we will not reckon that yielded by the first swarm, which is often more than that from the old stocks,) about one third of the average in good seasons. The second year there will be two to do the same; take this rate for ten years, we have 512 stocks, either of them worth as much as the non-swarmer, and about a thousand dollars worth of surplus honey. Call these stocks worth five dollars each, which makes $2,560, all added together will make the snug little sum of about $3,500, against $55. It is not to be expected that any of us will realize profits to this extent, but it is a forcible illustration of the advantages of the swarming hive over the non-swarmer.

PRINCIPLE OF SWARMING NOT UNDERSTOOD.

But many of these non-swarmers, 'tis said, can be changed to swarmers to suit the convenience of the apiarian—Colton's is one. It is asserted that it can be made to swarm within two days at any time, merely by taking off the six boxes or drawers that are very ingeniously attached; as this contracts the room, the bees are forced out. Now I will candidly confess that I could never get this thing to work at all. Of this I am quite positive, that he (Mr. Colton) is either ignorant of the necessary and regular preparations that bees make before swarming, or supposes others are. Mr. Weeks has advocated the same principle: he says, "There is no queen in any stage of existence, in the old stock, immediately after the first swarm leaves it." I have examined this matter till I am satisfied I risk but little in the bold assertion, that not one stock in fifty will cast a swarm short of a week after commencing preparations. This opinion will be adopted by whoever will take the trouble to investigate for themselves. (The chapter on swarming will give the necessary instructions for examining this point, if you wish.)

NOT TO BE DEPENDED UPON.

Further, these non-swarmers are not always to be depended upon as such. They will sometimes throw off swarms when there is abundant room in the hive as well as in the boxes.

HIVES NOT ALWAYS FULL BEFORE SWARMING.

I know Weeks, Colton, Miner and others, tell us the hive *must be full* before we need expect a swarm; but experience is against them. Bees do sometimes cast a swarm before filling the hive. From close observation, I find when a hive is very large, say 4,000 cubic inches, and is filled with comb, the first season, that such seldom swarm except in very good years.

SIZE OF HIVES NEEDED.

But if such hive is only half full, or 2,000 inches, it is very common for them to swarm without adding any new comb; proving very conclusively that a hive that size, is sufficient for all their wants in the breeding season. When about 1,200 inches only had been filled the first year, I have known them to add combs until they had filled about 1,800, and then cast a swarm, proving also that a little less than 2,000 will do for breeding. I have tested the principle of giving room to prevent swarming, a little further.

AN EXPERIMENT.

In the spring of '47, I placed under five full hives, containing 2,000 solid or cubic inches, as many empty ones, the same size, without the top. I had a swarm from each; but two had added any new comb, and these but little. If these hives had been filled to the bottom with comb in the spring, it is very doubtful whether either of them would have swarmed. The only place we can put a good stock and not expect it to swarm in good seasons, is inside a building, where it is perfectly dark, and even here a few have been known to do it. If we could manage to get *a very large hive* filled with combs, it would perhaps be as good a preventive as any. All the bees that could be reared in one season, would have sufficient room in the combs ready made for their labors, and there would be no necessity for their emigration. "But what becomes of all the bees raised in the course of several years?" To this question I shall not probably be able to give a satisfactory answer at present.

BEES DO NOT INCREASE, IF FULL, AFTER THE FIRST YEAR, IN SAME HIVE.

I only will notice the fact, that the bees somehow disappear, and there is no more at the end of five years than at the end of one. A stock of bees may contain 6,000 the first of May, and raise 20,000 in the course of the year; by the first of the next May, as a general thing, not one more will be found, even when no swarm had issued.

GILLMORE'S SYSTEM DOUBTED.

Now this fact is not known by a recent patentee from the State of Maine, (else he supposes others do not,) as he recommends placing bees in a house, and empty hives in connection with the one containing bees, and in a few years all will be full. He has discovered a mixture to feed bees, (to be noticed hereafter); this may account for an unusual quantity being stored by an ordinary sized family. He said another thing, that is, each of these added hives would contain a queen! This would seem to explain away the first difficulty of the continued increase of bees, and so it would if it did not get into another equally erroneous; one error never made another true. This idea of bees raising a queen, merely because they have a side box to the main hive, is contrary to all my experience, and to the experience of all writers (except himself) that I have consulted. If the principle is correct, why not sometimes raise a queen in a box on the top or side for us? I never discovered a single instance, where two perfect queens were quietly about their duties in connection with one hive. The deadly hostility of queens is known to all observing apiarians. Not having the least faith in the principle, I will leave it.

UTILITY OF MOTH-PROOF HIVES DOUBTED.

As for moth-proof hives, I have but little to say, as I have not the least faith in one of them. When I come to speak of that insect, I will show, I think, conclusively, that no place where bees are allowed to enter is safe from them.

Several other *perfect hives* might be mentioned; yet I believe that I have noticed the principles of each. Have I not said enough? Such as are not satisfied now would not be if I filled a volume. Our view of things is the result of a thousand various causes; the most powerful is interest, or prejudice.

It is said that in Europe, the same ingenuity is displayed in twisting and torturing the bee, to adapt her natural instinct to unnatural tenements; tenements invented not because the bee needs them, but because this is a means available for a little change. "Patent men" have found the people generally too ignorant of apiarian science. But let us hope that their days of prosperity in this line are about numbered.

INSTINCTS OF THE BEE ALWAYS THE SAME.

Let us fully understand that the nature of the bee, when viewed under any condition, climate, or circumstance, is the same. Instincts first implanted by the hand of the Creator, have passed through millions of generations, unimpaired, to the present day, and will continue unchanged through all future time, till the last bee passes from the earth. We may, we have, to gratify acquisitiveness, forced them to labor under every disadvantage; yes, we have compelled them to sacrifice their industry, prosperity, and even their lives have been yielded, but never their instincts. We may destroy life, but cannot improve or take from their nature. The laws that govern them are fixed and immutable as the Universe.

Spring returns to its annual task; dissolves the frost, warms into life nature's dormant powers. Flowers with a smile of joy, expand their delicate petals in grateful thanks, while the stamens sustain upon their tapering points the anthers covered with the fertilizing pollen, and the pistil springs from a cup of liquid nectar, imparting to each passing breeze delicious fragrance, inviting the bee as with a thousand tongues to the sumptuous banquet. She does not need an artificial stimulus from man, as an inducement to partake of the feast; without his aid or assistance she visits each wasting cup of sweetness, and secures the tiny drop, while the superabundant farina, dislodged from the nodding anthers, covers her body, to be brushed together and kneaded into bread. All she requires at the hands of man, is a suitable storehouse for her treasures. In good seasons, her nature Will prompt the gathering for her own use an over supply. This surplus man may appropriate to his own use, without detriment to his bees, providing his management is in accordance with their nature.

PROFIT THE OBJECT.

To give the bees all necessary advantages, and obtain the greatest possible amount of profit, with the least possible expense, has been my study for years. I might keep a few stocks for amusement, even if it was attended with no dollar and cent profit, but the number would be *very small*; I will honestly confess then, that *profit* is the actuating principle with me. I have a strong suspicion that the majority of readers have similar motives. I am sure, then, that all of us with these views, will consider it a pity, when a stock produces five dollars worth of surplus honey, to be obliged to pay three or four of it for patent and other useless fixings.

COMMON HIVE RECOMMENDED.

I would not exchange the hive I have used for the last ten years for any patent I ever saw, if furnished gratis. I will guarantee that it affords means to obtain surplus honey, as much in quantity and in any way which fancy may dictate, whether in wood or glass, and what is more than all, it shall cost nothing for the privilege of using.

SIZE IMPORTANT.

After deciding what kind of hive we want, the next important point is the size. Dr. Bevan, an English author, recommends a size "eleven and three-eighths inches square, by nine deep in the clear," making only about 1,200 inches, and so few pounds necessary to winter the bees, that when I read it, I found myself wondering if the English inch and pound were the same as ours.

SMALL HIVES MORE LIABLE TO ACCIDENTS.

At all events, I think it too small for our Yankee bees in any place. We must remember, that the queen needs room for all her eggs, and the bees need space to store their winter provisions; for reasons before given, this should be in one apartment. When this is too small, the consequences will be, their winter supply of food is liable to run out. The swarms from such will be smaller and the stock much more liable to accidents, which soon finish them off.

APT TO DECEIVE.

Yet I can imagine how one can be deceived by such a small hive, and recommend it strongly; especially if patented. Suppose you locate a large swarm in a hive near the size of Dr. Bevan's; the bees would occupy nearly all the room with brood-combs; now if you put on boxes, and as soon as filled put on empty ones, the amount of surplus honey would be great; very satisfactory for the first summer, but in a year or two your little hive is gone. This result will be in proportion as we enlarge our hives, until we arrive at the opposite extreme.

UNPROFITABLE IF TOO LARGE.

If too large, more honey will be stored than is required for their winter use. It is evident a portion might have been taken, if it had been stored in boxes. The swarms will not be proportionably large when they do issue, which is seldom—but there is this advantage, they last a long time, and are but little profit in surplus honey, or swarms.

CORRECT SIZE BETWEEN TWO EXTREMES.

Between the two extremes, like most other cases, is found the correct place. A hive twelve inches square, each way, inside, has been recommended as the correct size. Here are 1,728 cubic inches. This, I think, is sufficient for many places, as the queen probably has all the room necessary for depositing her eggs; and as the swarms are more numerous, and nearly as large as from hives much larger; also, there is room for honey sufficient to carry the bees through the winter, at least, in many sections south of 40 degrees latitude, where the winter is somewhat short.

SIZE FOR WARM LATITUDES.

This size will also do in this latitude (42 degrees,) in some seasons, but not at all in others.[3] Not one swarm in fifty will consume twenty-five lbs. of honey through the winter, that is, from the last of *September* to the first of April, (six months). The average loss in that time is about eighteen lbs.; but the critical time is later; about the last of May, or first of June, in many places.

LARGER HIVE MORE SAFE FOR LONG WINTERS OR BACKWARD SPRING.

About the first of April they commence collecting pollen and rearing their young; by the middle of May all good stocks will occupy nearly, if not quite all, their brood-combs for that purpose, but *little honey is obtained* before fruit blossoms appear; when these are gone, no more of any amount is obtained until clover appears, which is some ten days later. (I am speaking now particularly of this section; I am aware it is very different in other places, where different flowers exist.) Now if this season of fruit flowers should be accompanied by high winds, or cold rainy weather, but little honey is obtained; and our bees have a numerous brood on hand that *must be fed*. In this emergency, if no honey is on hand of the previous year, a famine ensues; they destroy their drones, perhaps some of their brood, and for aught I know put the old bees on short allowance. This I do know, that the whole family has actually starved at this season; sometimes in small hives. This of course depends on the season; when favorable, nothing of the kind occurs. Prudence therefore dictates the necessity of a provision for this emergency, by making the hive a little larger for northern latitudes, as a little more honey will be stored to take them through this critical period. From a series of experiments closely observed.

2,000 INCHES SAFE FOR THIS SECTION.

I am satisfied that 2,000 inches in the clear, is the proper size for safety in this section, and consequently, profit. On an average, swarms from this size are as large as any.

The dimensions should be uniform in all cases, whatever size is decided on. It is folly to accommodate each swarm with a hive corresponding in size; a very small family this year, may be very large next, and a very large one, very small, &c. A queen belonging to a small swarm will be capable of depositing as many eggs, as another belonging to a barrel full. A small family able to get through the winter and spring, may be expected by another year to be as numerous as any.

KIND OF WOOD, WIDTH OF BOARD, ETC.

Of the kinds of wood for hives, pine is preferable, still other kinds will do; I have no faith in bees liking one kind better than another, and less likely to leave on that account. Hemlock is cheaper, and used to a great extent; when *perfectly sound* is as good as anything, but is very liable to split, even after the bees have been in them some time. It should be used only when better wood cannot be obtained. Bass wood when used for hives should *always be painted*, and then will be very liable to warp from the moisture arising from the bees inside. When not painted outside, and allowed to get wet, if only for a few hours, so much moisture is absorbed that it will bend outward, and cleave from the combs and crack them. A few days of dry weather will relieve the outside of water, and the inside kept moist by the bees, the bending will be reversed, and the combs pressed inward, keeping the bees fixing that which will not "stay fixed." Perhaps there is wood as suitable or better than pine, but it is not as common.

SHAPE OF NO CONSEQUENCE.

Boards should be selected, if possible, that will be the proper width to make the hive about square, of the right size. Say twelve inches square, inside, by fourteen deep. I prefer this shape to any other, yet it is not all important. I have had some ten inches square by twenty in length; they were awkward looking, but that was all, I could discover no difference in their prosperity. Also, I have had them twelve inches deep by thirteen square, with the same result. Hence, if we avoid extremes, and give the required room, the shape can make but little difference.

It has been recommended to plane the boards for hives, "inside and out;" but bees, when first put into such hive, find much difficulty in holding fast until they get their combs started, hence this trouble is worse than useless.

DIRECTIONS FOR MAKING HIVES.

If hives are not desired of the cheapest possible construction, the outside may be planed and painted; but it is doubtful whether strict economy would demand it. Yet a painted hive appears so much better, that it ought to be done, especially as the paint adds almost enough to its durability to pay the expense. The color may be whatever fancy dictates; the moth will not probably be attracted by one color more than another. White is affected the least by the sun in hot weather. Lime is put on as white-wash, annually, by many, as a protection against insects.

When hives are not painted, the grain should never be crosswise, having the width of boards form the height; not that the bees would have any dislike to such, but nails will not hold firmly, they draw out in a few years. The size, shape, materials, and manner of putting together, are now sufficiently understood, for what I want. Sticks half an inch in diameter, should cross each way through the centre, to help support the combs. A hole about an inch diameter in the front side, half way to the top, is a great convenience for the bees to enter when coming home heavy laden.

It now remains to make the top, cover, and boxes, (the bottom-board will be described in another chapter.) The tops should be all alike; boards fifteen inches square are just the right size; three-fourths of an inch is the best thickness, (inch will do;) plane the upper side, rabbet out around the edge of the upper side one inch wide, and three-eighths deep; this will leave the top inside the rabbeting, just thirteen inches.

SIZE OF CAP AND BOXES.

A box for a cover or cap, that size inside, will fit any hive. The height of this box should be seven inches. Of course other sizes will do, but it is best to commence with one that we can adhere to uniformly, and no vexations arise by covers not fitting exactly, &c. I think this size is as near correct as we shall be likely to get; we want all the room in the boxes that the majority of our stocks demand for storing in a yield of honey,[4] at the same time not be necessitated to give too much of the room in the height. They will commence work in a box five inches high, much sooner than one seven or eight. To give the requisite room, and have the boxes less than five inches high, would require more than thirteen inches on the top, this would make the hive too much out of shape; it would appear top-heavy.

MINER'S HIVE.

Miner's Equilateral Hive has a cap somewhat smaller than this in diameter; consequently, if we have the requisite room, it must be in its height. But by making the cap of his a little larger, and a few trifling alterations, it would do very well for a patent. And if any one *must* have a patent hive, my advice is to get that; it costs but two dollars for the right of using, and is nearer what we want for bees, than any I ever saw. I prefer rabbeting around the edge of the top, instead of nailing on a thin board the size of the inside of the cover, with room for a slide under it; it affords too nice a place for worms to spin their cocoons. Also, without the rabbeting water may get under the cap, and pass along the top till a hole lets it among the bees. As for slides, I do not approve of them at all; in shutting off communication, it is almost certain to crush a few bees. This makes them irritable for a week; they are unnecessary for me, at least. We will now finish the hive.

DIRECTIONS FOR MAKING HOLES.

After the top is got out as directed, strike a line through the centre, three and a quarter inches from this, make another on each side, now measure on one of the last lines, two and a half inches for the first hole, two inches for the next, and so on till five are marked on this, and the same number on the other side, ten in all; these holes should be about an inch diameter, a pattern three and a quarter inches wide, and thirteen in length, with places for holes marked on it, will save time when many are made. When this top is nailed on, the hive is ready. A less number of holes is often used, and one is thought by some to be sufficient; experience has satisfied me that the more room bees have to enter boxes, the less reluctance is manifested in commencing their work in them; but here is another extreme to be avoided: when the holes are much larger, or more of them, or even one very large one, the queen is very apt to go into the boxes and deposit her eggs, which renders the comb tough, dark, &c., also bee-bread is stored near the brood. Dr. Bevan's and Miner's cross-bar hives are objectionable on this account, they offer too free access to the boxes; we want all the room that will answer, and no more.

A SUGGESTION.

Mr. Miner's cross-bar hive is intended to make the bees construct all straight combs, and probably will do it. But the disadvantage of bee-bread and brood in the boxes will not be made up by straight combs.

For the benefit of those who have been made to believe straight combs *all important*, and perhaps have purchased the right to make the hive, and had some constructed, and have found bee-bread in their surplus honey, I would suggest an improvement, (that is, if it is thought the straight combs will pay. If you have not the right for the cross-bar hive, and you wish to use it, I would say, buy the right, and remove all grounds of complaint with him.) Put in the bars and hive your bees as he directs. After all the combs are started, instead of setting the open bottom boxes (which are also unsuitable for sending to market) directly on the bars as he recommends, take off the cloth, and with screws fasten on a top with ten holes, that I have just described; and then you will have the straight combs, and surplus honey in the boxes pure.

GLASS BOXES PREFERRED.

Having told how I make a hive, I will now give some reasons for preferring a particular kind of boxes. I have taken great quantities of honey to market, put up in every style, such as tumblers, glass jars, glass boxes, wooden boxes with glass ends, and boxes all wood. I have found the square glass boxes the most profitable; the honey in such appears to the best possible advantage, so much so, that the majority of purchasers prefer paying for the box at the same rate as the honey, than the wood box, and have the tare allowed. This rate of selling boxes always pays the cost, while we get nothing for the wood. Another advantage in this kind of boxes is, while being filled, the progress can be watched, and the time they are finished known precisely, when they should be taken off, as every day they remain after that, soils the purity of the combs.

GLASS BOXES—HOW TO MAKE.

Directions for making.—Select half-inch boards of pine or other soft light wood, cut the length twelve and three-quarters inches, width six and three-eighths inches, dress down the thickness to three-eighths or less, two pieces for a box, top and bottom, in the bottom bore five holes throughout the centre to match with those in the top of the hive, (the pattern used in marking the top of hives is just the one to mark these). Next, get out the corner posts, five-eighths of an inch square, and five inches in length; with a saw, thick enough to fit the glass, cut a channel length-wise on two sides, one-fourth of an inch deep, one-eighth from the corner, for the glass. A small lath nail through each corner of the bottom into the posts will hold them; it is now ready for the glass—10×12 is the right size to get—have them cut through the centre the longest way for the sides, and they are right, and again the other way, five and five-eighths long for the ends. These can now be slipped into the channels of the posts, and the top nailed on like the bottom, and the box is ready.

GUIDE-COMBS NECESSARY.

It will be found a great advantage, previous to nailing on the top, to stick fast to it some pieces of guide-combs in the direction you wish the bees to work. They are also an inducement for them to commence several days sooner, than if they had to start combs for themselves;[5] a piece an inch square will do; it is well to start every comb you want in the box; two inches apart is about the right distance to look well. To make these pieces hold fast, melt one edge by the fire, or candle, or melt some bees-wax, and dip one edge in that, and apply it before cooling; with a little practice you can make them stick without difficulty. For a supply of such combs, save all empty, clean, white pieces you can, when removing combs from a hive.

If you have any way superior to this for making glass boxes, so much the better, make them so by all means: "The best way is as good as any." I give my method to be used only when better is not convenient. If you sell honey, I think you will find it an advantage to have glass boxes made in some way. Two of this size when full weigh 25 lbs. If preferred, four boxes six and three-eighths inches square, can be used for a hive instead of two; the expense of making is a little more for the same number of lbs., yet, when it is in market, a few customers will prefer this size.

WOOD BOXES.

For home consumption, the wood-box will answer equally well for all purposes of obtaining the honey, but will give no chance to watch the progress of the bees, unless a glass is inserted for the purpose, and then it will need a door to keep it dark, or a cover over the whole like the one for glass boxes, may be put on. Wood boxes are generally made with open bottom, and set on the top of the hive. A passage for the bees out of the box to the open air is unnecessary, and worse than useless. They like to store their honey as far from the entrance as possible. Unless crowded for room, they will not store much there when such entrances are made.

Whether we intend to consume our surplus honey or not, it is as well to have the hives and covers made in a manner that we can use glass, when we are likely to have some to spare. I am not sure, but it would pay to make hives in this way, even if glass boxes were never used; the rabbeting prevents light as well as water from passing under the cover; imagine a box set on a plain board nailed on for a top, without the rabbeting; the warping or bending admits the light and water, especially when hives are out in the weather, (and I shall not recommend any other way of keeping them.)

COVER FOR HIVES.

I have termed the cap or box a cover; but this should also be covered with a board laid on, if nothing else. A good roof for each hive can be made by fastening two boards together like the roof of a building; let it be about 18 by 24 inches; it being loose, can be changed in accordance with the season; in spring, let the sun strike the hive; but in hot weather let the longest end project over the south side, &c. You can ornament this hive, if you choose, by mouldings or dentals, under the top, where it projects over the body of the hive, also the cap can have the top projected a little and receive the same addition.

JARS AND TUMBLERS—HOW PREPARED.

When jars, tumblers, or other vessels, that are all glass, are used, it is *absolutely necessary* to fasten as many pieces of combs as you wish made, in the top, for a beginning, or fasten a piece of wood there; as they seldom commence building on glass, without a start.

Some of you may have seen paraded at our fairs, or in the public parts of some of our cities, hives containing tumblers, some of them neatly filled, others empty, and this meagre sentence written upon them, *not to be filled!* Pretending to govern the bees, as the juggler sometimes does his tricks, by mysterious incantations! I once encountered an agent of this humbug, and modestly suggested to him that I had a counter charm: that I could put a tumbler on his hive and it would be filled if the others were, however much he might forbid it by written charms! He saw at a glance how the matter stood; I was not the customer he wanted, and intimated that the show was only intended for the extreme verdancy of most visitors. It no doubt assisted in displaying his profound knowledge in bee management, which he wished to establish, as he had a little work on the subject to sell, also hives, and bees. The reader no doubt will guess as I did, the reason that those tumblers were not filled, was because no combs were put in for a start.

PERFECT OBSERVATORY HIVE DESCRIBED.

There are many things pertaining to bees that cannot be properly examined and understood, without a glass hive of some sort. Yet a perfect observatory hive containing but one comb, is not a perfect hive for the bees. We can see very well what the bees are doing, but it is not a tenement they would choose if left to themselves. It forces them to labor in an unnatural manner, is unsuitable for wintering bees, and otherwise but little profit. If the satisfaction of witnessing some of their operations more perfectly than in glass hives of another kind will not pay, it is doubtful if we get it. I will describe as briefly as possible. Two frames or sashes about two and a half feet square, containing glass, are so fastened together as to leave room for only one comb between them, about an inch and three-fourths apart. A comb of this size will not support itself by the top and edges; hence, it is necessary to put in numerous cross-bars to assist in supporting it. Outside the glass are doors to keep the whole dark, to be opened when we wish to inspect proceedings. Under the bottom is a board or frame, to keep it in an upright position, &c. Probably but few will be induced to make one. I will therefore describe another; a hive that I think will pay better.

ONE LIKE COMMON HIVE PREFERRED.

If we expect to know what bees are doing in ordinary hives, we must have one similar in every respect, in size, shape, number of bees, &c. The construction of royal cells will be watched by most observers with the greatest interest; now these are generally on one edge of the combs. The bees leave a space half an inch or more between the edges of the combs and one side of the hive, near half the length of it, apparently for no other purpose but to have room for these cells, as the other edges of the same combs are generally attached to the hive at the bottom.

WHAT MAY BE SEEN.

Now instead of having one piece or pane of glass in the side of several hives, I would recommend having one or more with glass on every side; because we might have it on three sides, and not the fourth; and this might contain all the queen cells, and we should miss an important sight. There are many other things to be witnessed in such a hive. The queen may be often seen depositing her eggs! We may see the workers detach the scales of wax from their abdomen, and apply them to the combs during the process of construction, see them deposit pollen from their legs, store their honey, feed the queen, each other, their young brood, seal over cells containing brood, honey, &c. It is further useful as a guide for putting boxes on other hives, (that is, if it is a good one, which it should be); we can easily ascertain whether our bees are gaining or losing.

DIRECTIONS FOR MAKING GLASS HIVE.

My method of making them is as follows: The top is like those for other hives, fifteen inches square, adapted to boxes and cover. This hive we want to be as profitable as any, giving us surplus honey, and swarms like others. Four posts are then got out, two inches square, and thirteen in length; care should be taken to have the ends perfectly square.

A frame is then to be made, just fourteen inches square outside, for the bottom; the pieces are one inch thick, by two in width, halved together at the corners. A guage-mark is then made around the under side of the top, half an inch from the edge, a post is then set inside of each corner of this mark, and thoroughly nailed, the bottom is nailed on with the posts even with the outside corners. Four pieces an inch thick, and an inch and a half wide, are fitted between the posts, even with the guage-mark on the top. Sixteen strips, about one quarter by half an inch, are got out, eight to be ten, and eight twelve inches long.

A gauge-mark one inch from posts, bottom, &c., is the place to nail these strips; very small nails or tacks will hold them. The panes of glass are to rest against them, which are held in their places by small pieces of tin, or brads. The doors are the size of the glass, 10×12, about three-fourths of an inch thick; these doors are cut a little too short, and the pieces, to prevent warping, are nailed on the ends; these are hung to a post on one side, and secured by a button on the other. On two opposite sides inside the posts, half way up, two strips, half an inch by three quarters, are nailed, with holes in them for the cross-sticks; one way is enough if you have guide-combs for a start, like those recommended for boxes, so that the sheets will be at right angles with them; otherwise, let the sticks cross both ways, about three each way will be needed, as the glass at the edges is not so good a support as wood.

The cap can be made of half inch boards; the top to project over like the hive, or let it be a little more than half an inch, it will admit a heavier moulding, which should surround it here, as well as at the top of the hive, or if it is prefered, dentals can be used, and look equally well—when no ornament is wanted, omit it. But painting seems necessary for such hives, to prevent warping, and the swelling of the doors in wet weather; these want to open and shut without rubbing or sticking, otherwise we disturb the bees every time a door is stirred. Putty should not be used to hold the glass, as the bees in the course of a few years will cover it with propolis; it is then necessary to take it out, and scrape, clean, and return it, when, if fastened with putty, it would be difficult; cold weather is the time for this operation. I am aware that a hive can be more substantially made than the one here described; but I have endeavored to make one as cheap as possible, and if properly made, will answer. The cost will be much less than many patents, and the satisfaction much more, at least, with many. When our hive contains a swarm of bees, and they are thoroughly in operation, we must not let them pass out at the bottom on every side, as they are frequently allowed to do from other hives; because, should one come out a little excited in consequence of a slight jar, accidentally given the hive, on opening the door or some other way, and should find our face within a foot of their house, peering in the window among their works, it would be very likely to give us *a gentle hint* that it was a mark of low breeding, that we were not wanted there at all, and that it was none of our business what they were doing. To prevent this as far as possible, a bottom-board, somewhat different from the common one, is needed. Four posts of chestnut or other lasting wood, about two inches square, are driven into the earth in the form of a square, far enough apart to come under the corners of the bottom-board, (fifteen inches,) and high enough for convenience when looking into the hive. The ends of these posts are to be perfectly level, and to which the bottom is to be nailed fast. As the hive is to sit perfectly close to the board, a passage must be made through it, as well as means for ventilation in hot weather, without raising the hive for that purpose. It requires a board about fifteen inches square, planed smooth, the ends clamped to prevent warping or splitting; a portion of the centre is taken out, say six inches by ten, and wire cloth nailed over, four-ounce tacks will hold it, fasten it just enough to keep the bees from getting through; very likely it will want to be taken off occasionally and cleaned from the propolis that will be spread over it. It is easiest done in freezing weather.

Take an edge in each hand, and rock the wires a few times out of square, and it will readily crumble and fall out. In warm weather it must be scalded or burnt off. To close this space, a moving slide is fixed in grooves under-side, fastened to the posts or board. The slide is to be moved in accordance with the weather, when cold, close it, when hot, withdraw it, and give the bees as much air as possible, without raising the hive, the whole of such space is as much ventilation as ordinary hives raised an inch. (Wire cloth is needed for other purposes, it is best to procure some, even at considerable trouble and expense.) On the side of the board intended for the front, two inches from the edge of the wire cloth, a passage is cut for the bees, three-eights of an inch wide, by eleven in length. "But how is the bees to get to this place, so inconvenient, something is needed to assist them?" Certainly, Sir; an alighting board, eleven inches wide, and about two feet long, (not planed), is placed at an angle of forty-five degrees, between the two front posts of your stand, the upper end passing under the bottom, far enough back; to be just even with the back-side of the passage for the bees. The bees alight on this board, and walk up into the hive without difficulty. When the bees are at work pretty freely, and a door of this hive is opened, those that are about departing will be very likely to get on the glass, instead of through the opening at the bottom; seeing the light through the glass, they endeavor to escape by the nearest route. When so many gather here as to prevent a good view, and you wish to observe further, shut the door a moment and they will leave through their own passage, when you can open your door again, for a short time.

After the hive is filled with combs, the number attracted to the glass on opening a door will be much less.

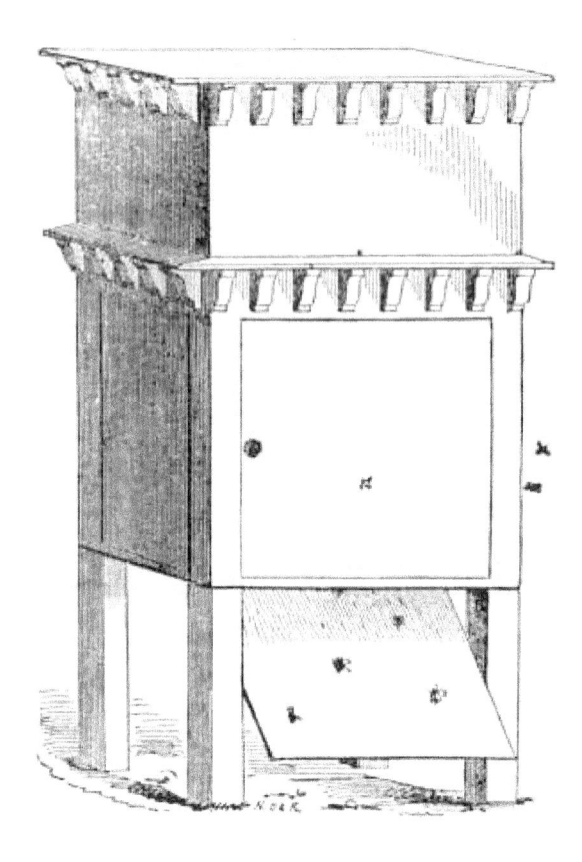

The plate on the preceding page represents a glass hive, cover, and stand. The common hive can be made equally ornamental, if you choose; this kind of stand is unnecessary for them. I use such as are recommended on page 138.

CHAPTER III.

BREEDING.
IMPERFECTLY UNDERSTOOD.

The time that bees commence raising their young brood is but imperfectly understood by most people. Many persons that have kept them for years, have bestowed so little attention on this point, that they are unable to tell at what time they commence, how they progress, or when they cease. A kind of an idea that one swarm, and occasionally two or three, are reared sometime in June, or fore part of summer, is about the extent of their reflections on the subject. Whether the drones deposit the eggs, or that a portion of the workers are females, and each raise a young one or two, or whether the "king bee" is the chap for laying eggs, is a matter beyond their ability to answer. It is but a few years since, that a correspondent of a Journal of Agriculture denied the existence of a queen bee, giving the best reasons he had, no doubt, that is, he had never seen one. But bee-keepers of this class are so few, it is unnecessary to waste time to convince them; suffice it to say, that a queen exists with every prosperous swarm, and all apiarians with much pretensions to science, acknowledge the fact, also, that she is the mother of the whole family.

The period at which they commence depositing eggs probably depends on the strength of the colony, amount of honey on hand, &c., and not the time they commence gathering food.

GOOD STOCK SELDOM WITHOUT BROOD.

I once removed the bees from a hive on the tenth of January, and found brood amounting to about five hundred, sealed over, and others in every stage of growth down to the egg.

This hive had been in the house, and kept warm; it will doubtless be supposed that being kept warm was the cause; but this is not a solitary instance. A neighbor lost a hive the fourteenth February, in weather cold enough to seal the entrance with ice, and smother the bees. I assisted to remove the combs, and found young brood in abundance, from the perfect bee, through all stages of growth. This stock had been in the cold all winter. I have further noticed, when sweeping out the litter under the hives early in spring, say the first of March, that young bees would often be found under the best stocks. Hence it appears there is but little time, and perhaps none, when our best stocks have no broods. Yet stocks, when very weak, do not commence till warm weather. It seems that a certain degree of warmth is necessary to perfect the brood, which a small family cannot generate.

HOW SMALL STOCKS COMMENCE.

The first eggs are deposited in the centre of the cluster of bees, in a small family; it may not be in the centre of the hive in *all* cases; but the middle of the cluster is the warmest place, wherever located. Here the queen will first commence; a few cells, or a space not larger than a dollar, is first used, those exactly opposite on the same comb are next occupied. If the warmth of the hive will allow, whether mild weather produces it, or the family be large enough to generate that which is artificial, appears to make no difference; she will then take the next combs exactly corresponding with the first commencement but not quite as large a place is used as in the first comb. The circle of eggs in the first is then enlarged, and more are added in the next, &c., continuing to spread to the next combs, keeping the distance to the outside of the circle of eggs, to the centre or place of beginning, about equal on all sides, until they occupy the outside comb. Long before the outside comb is occupied, the first eggs deposited are matured, and the queen will return to the centre, and use these cells again, but is not so particular this time to fill so many in such exact order as at first. This is the general process of small or medium sized families. I have removed the bees from such, in all stages of breeding, and always found their proceedings as described.

DIFFERENT WITH LARGER ONES.

But with very large families, their proceedings are different: as any part of the cluster of bees is warm enough for breeding, there is less necessity for economizing heat, and having all the eggs confined to one small spot, some unoccupied cells will be found among the brood; a few will contain honey and bee-bread.

HOW POLLEN IS STORED IN THE BREEDING SEASON.

But in the height of the breeding season, a circle of cells nearly all bee-bread, an inch or two wide, will border the sheets of comb containing brood. As bee-bread is probably the principal food of the young bee, it is thus very convenient.

When pollen is abundant, and the swarm is in prosperous condition, they soon reach the outside sheets of comb with the brood. At this period, when the hive is about full, and the queen is forced to the outside combs to find a place for her eggs, it is interesting to witness operations in a glass hive. I have seen her several times during one day, on the same piece of comb (next the glass). The light has no immediate effect on her "Highness," as she will quietly continue about her duty, not the least embarrassed by curious eyes at the window. Before depositing an egg, she enters the cell head first, probably to ascertain if it is in proper condition to receive it; as a cell part filled with bee-bread or honey is never used. If the area of combs is small, or the family is small, and cannot protect a large space with the necessary heat, she will often deposit two, and sometimes three, in one cell (the supernumeraries I suppose are removed by the workers). But under prosperous circumstances, with a hive of suitable size, &c., this emergency is avoided.

OPERATION OF LAYING AND THE EGGS DESCRIBED.

When a cell is in a condition to receive the egg, on withdrawing her head she immediately curves her abdomen, and inserts it a few seconds. After leaving it, an egg may be seen attached by one end to the bottom; about the sixteenth of an inch in length, slightly curved, very small, nearly uniform the whole length, abruptly rounded at the ends, semi-transparent, and covered with a very thin and extremely delicate coat, often breaking with the slightest touch.

After the egg has been about three days in the cell, a small white worm may be seen coiled in the bottom, surrounded with a milky-like substance, which is its food, without doubt. How this food is prepared, is merely guess-work. The hypothesis of its being chiefly composed of pollen, I have no objection to; as it is sufficiently proved by the quantities that accumulate in hives that lose their queen, and rear no brood (that is, when a requisite number of workers are so left). The workers may be seen entering the cell every few minutes, probably, to supply this food.[6]

TIME FROM THE EGG TO THE PERFECT BEE.

In about six days it is sealed over with a convex waxen lid. It is now hidden from our sight for about twelve days, when it bites off the cover, and comes forth a perfect bee. The period from the egg to the perfect bee varies from twenty to twenty-four days; average about twenty-two for workers, twenty-four for drones. The temperature of the hive will vary some with the atmosphere; it is also governed by the number of bees. A low temperature probably retards the development, while a high one facilitates it. You may have seen accounts of the assiduous attentions given to the young bee when it first emerges from the cell: 'tis said they "lick it all over, feed it with honey," &c., desperately pleased with their new acquisition.

ROUGH TREATMENT OF THE YOUNG BEE.

Now, if you expect to see anything of this, you must watch a little closer than I have. I have seen hundreds when biting their way out. Instead of care or notice, they often receive rather rough treatment: the workers, intent on other matters, will sometimes come in contact with one part way out the cell, with force sufficient to almost dislocate its neck; yet they do not stop to see if any harm is done, or beg pardon. The little sufferer, after this rude lesson, scrambles back as soon as possible out of the way; enlarges the prison door a little, and attempts again, with perhaps the same success: a dozen trials are often made before they succeed. When it does actually leave, it seems like a stranger in a multitude, with no friend to counsel, or mother to direct. It wanders about uncared for and unheeded, and rarely finds one sufficiently benevolent to bestow even the necessaries of life; but does sometimes. It is *generally* forced to learn the important lesson of looking out for itself, the day it leaves the cradle. A cell containing honey is sought for, where its immediate wants are all supplied.

GUESS WORK.

The time before it is ready to leave the hive for honey, I might guess would be two or three days. Others have said "it would leave *the day it left the cell*;" but I guess they guess at this point. They tell us, too, that after the bees seal over the cells containing the larvæ, "they immediately commence spinning their cocoons, which takes just about thirty-six hours." I think it very likely; but when I admit it, I cannot imagine how it was ascertained;—the faculty of looking through a mill-stone I do not possess, and it requires about the same optical penetration to look into one of these cells after it is sealed over, as it is all perfect darkness. Suppose we drive away the bees and open the cell, to give us a look at the interior: the little insect stops its labor in a moment, probably from the effect of air and light. I never could detect one in its labor. Suppose we open these cells every hour after sealing; can we tell anything about their progress by the appearance of these cocoons, or even tell when they are finished? The thickness of a dozen would not exceed common writing paper. When a subject is obscure, or difficult to ascertain, like this, why not tell us how they found out the particulars; and if they were guessed at, be honest, and say so? When the bee leaves the cell, a cocoon remains, and that is about all we *know* about it.

TERMS APPLIED TO YOUNG BEES.

The young bee, when it first leaves the egg, is termed grub, maggot, worm, or larva; from this state it changes to the shape of the perfect bee, which is said to be three days after finishing the cocoon; from the time of this change, till it is ready to leave the cell, the terms nymph, pupa, and chrysalis, are applied. The lid of the drone's cell is rather more convex than that of the worker's, and when removed by the young bee to work its way out, is left nearly perfect; being cut off around the edges, a good coat or lining of silk keeps it whole; while the covering of the worker's cell is mostly wax, and is pretty well cut to pieces by the time the bee gets out. The covering to the queen's cell is like the drone's, but larger in diameter, and thicker, being lined with a little more silk.

DISCREPANCY IN TIME IN REARING BROOD AS GIVEN BY HUBER.

We are told by most writers, the period of time necessary to perfect from the egg, the three different kinds of bees. Huber leads the way, and the rest, *supposing him to be right,* repeat in substance his account as follows: That the whole time necessary to perfect a queen from the egg is sixteen days, the worker twenty, and the drone twenty-four days; Huber (as quoted by Harpers) gives the time of each stage of development belonging to each kind of bee; but is rather unfortunate in arithmetic; the items, or stages, when added together, "do not prove," as the school-boys say; that is, he gains time by making his bee by degrees. He says, first, of the worker, "It remains three days in the egg, five in the grub state, it is thirty-six hours in spinning its cocoon; in three days it changes to a nymph, passes six in that form, and then comes forth a perfect bee." How do the items add?

The egg, 3 days. Grub, 5 " Spinning cocoon, 1-1/2 " Changing to a nymph, 3 " In that form, 6 " ———— 18-1/2 days.

One and a half days short. We will next see how the figures with the royal insect match; recollect sixteen days are all she has allowed; then, of the different stages, "three days in the egg, is five a worm, when the bees close its cell, and it immediately begins its cocoon, which is finished in twenty-four hours. During eleven days, and even sixteen hours of the twelfth, it remains in a state of complete repose. Its transformation into a nymph then takes place, in which state four days and part of the fifth are passed." Now let us add the items:

The egg, 3 days. A worm, 5 " Spinning a cocoon, (24 hours), 1 " Reposes eleven days and 16 hours, 11-2/3 " A nymph four days, and part of the fifth, 4-1/3 " ———— 25 days.

Now, reader, what do you make of such palpable blundering guess-work? A difference of nine days—the merest school-boy ought to know better! Can we rely on such history? Does it not prove the necessity of going over the whole ground, applying a test to every assertion, and a revision of the whole matter throughout? My object is not to find fault, but to get at *facts*. When I see such guess-work as the above published to the world, in this enlightened age, gravely told to the rising generation, as a portion of natural history, I feel it a duty not to resist the inclination to expose the absurdity.

THE NUMBER OF EGGS DEPOSITED BY THE QUEEN GUESSED AT.

The number of eggs that a queen will deposit is often another point of guess-work. When the estimate does not exceed 200 per diem, I have no reason to dispute it; the number will probably fall short in some cases, and exceed it in others. Some writers suppose that this number "would never produce a swarm, as the bees that are lost daily amount to, or even exceed that number," and give us instead from eight hundred to four thousand eggs in a day, from one queen. The only way to test the matter accurately, is by actually counting, in an observatory hive, or in one with sufficient empty combs to hold *all the eggs* she will deposit for a few days, when, by removing the bees, and counting carefully, we might ascertain, and yet several would have to be examined, before we could get at the average. The nearest I ever came to knowing anything about it happened as follows: A swarm left, and the queen from some cause was unable to cluster with it, and was found, after some trouble, in the grass a few rods off. She was put in the hive with the swarm about 11 o'clock, A.M.; the next morning, at sunrise, I found on the bottom-board, among the scales of wax, 118 eggs that had been discharged in that time. Probably a few escaped notice, as the color is the same as wax scales; also, they might already have had combs containing some. I have several times found a few the next morning, under swarms hived the day previous, but never over thirty, except in this one instance. The reason of this queen not being able to fly well might have been an unusual burden of eggs. Perhaps it would be as well to mention here, that in all cases where eggs are found in this way, that they must be first swarms which are accompanied by the old queens.

Schirach estimates "the eggs a single female will lay, from 70,000 to 100,000 in a season." Reaumer and Huber do not estimate so high. Another writer estimates 90,000, in three months. Let the number be as it may, probably thousands are never perfected. During the spring months, in medium and small families, where the bees can protect with animal heat but a few combs, I have often found cells containing a plurality of eggs, two, three, and occasionally four, in a single cell. These supernumeraries must be removed, and frequently may be found amongst the dust on the bottom-board.

A TEST FOR THE PRESENCE OF A QUEEN.

If you have a hive that you suspect has lost a queen at this season, her presence can be ascertained nine times in ten by this method. Sweep off the board clean, and look the next day or two after for these eggs. Take care that ants, or mice, have no chance to get them; they might deceive you, being as fond of eggs for breakfast as anyone.[7] When one or more is found, or any immature bees, it is sufficient, no further proof of the presence of a queen is needed.

Another portion of eggs is wasted whenever a supply of their food fails; if we remove the bees from a stock during a scarcity, when the hive is light, we will be very likely to find hundreds of eggs in the cells, and but very few advancing from that stage towards maturity. I have thus found it in the fall, in July, and sometimes the first of June, or at any time when maturing the brood would be likely to exhaust their stores, to endanger the family's supply. Now, instead of the fertility of the queen being greater in spring and first of summer than at other times, (as we are often told), I would suggest the probability that a greater abundance of food at this season, and a greater number of empty cells, may be the reason of the greater number of bees matured.

WHEN DRONES ARE REARED.

Whenever the hive is well supplied with honey, and plenty of bees, a portion of eggs are deposited in the drone-cells, which three or four days more are necessary to mature than the worker.

WHEN QUEENS ARE REARED.

Also, when the combs become crowded with bees, and honey plenty, the preparations for young queens commence: as the first step towards swarming, from one to twenty royal cells are begun; when about half completed, the queen (if all continues favorable) will deposit eggs in them, these will be glued fast by one end like those for the workers; there is no doubt but they are precisely the same kind of eggs that produce other bees. When hatched, the little worm will be supplied with a superabundance of food; at least, it appears so from the fact, that a few times I have found a quantity remaining in the cell after the queen had left. The consistence of this food is about like cream, the color some lighter, or just tinged with yellow. If it was thin like water, or even honey, I cannot imagine how it could be made to stay in the upper end of an inverted cell of that size in such quantities as are put in, as the bees often fill it near half full. Sometimes a cell of this kind will contain this food, and no worm to feed upon it. I *guessed* the bees had compounded more than their present necessities required, and that they stored it there to have it ready, also, that being there all might know it was for royalty.

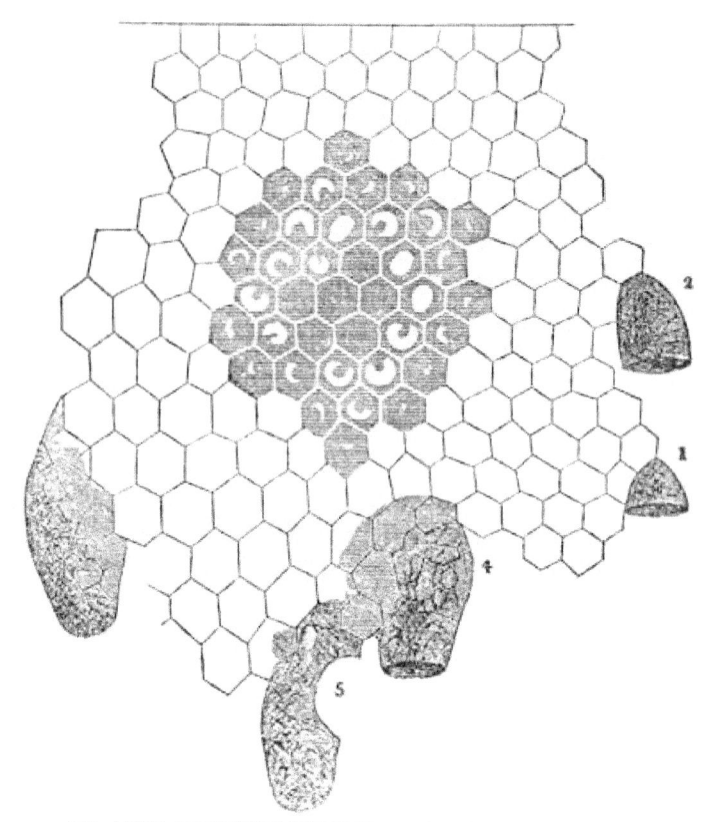

PLATE OF THE THREE KINDS OF CELLS.

The taste is said to be "more pungent" than food given to the worker, and the difference in food changes the bee from a worker to a queen. I have nothing to say against this hypothesis; it may be so, or the young bee being obliged to stand on its head may effect it, or both causes combined may effect the change. I never tasted this food, or found any test to apply.

The preceding plate represents a piece of comb containing all the different cells—those at the left hand the size for drones. In the centre are few that appear sealed over, others nearly covered, others the larva in different stages of growth, as well as the eggs. *Fig. 1* represents a queen's cell just commenced. They are usually started thus far the first season, very frequently when the hive is only half or two-thirds full. *Fig. 2* is a cell sufficiently advanced to receive the egg. *Fig. 3* one finished, the stage when the first swarm leaves. *Fig. 4* when a queen has been perfected and left. *Fig. 5* is a cell where its occupant has been destroyed by a rival, and removed by the workers. It will be perceived that each finished queen's cell contains as much wax as fifty made for the workers.

LIABILITY OF BEING DESTROYED.

In any stage from the egg to maturity these royal insects are liable to be destroyed;—if honey fails from any cause sufficient to make the existence of a swarm any way hazardous, the preparations are abandoned, and these young queens destroyed; (I would here request the reader not to condemn me for telling more than I can prove, until he has had the whole story; in the swarming season, I will give further particulars.)

DRONES DESTROYED WHEN HONEY IS SCARCE.

When an occurrence like the above happens, the drones next fall victims to the failure of honey. A brief existence only is theirs; such as are perfect, are destroyed without mercy; those in the chrysalis state are often dragged out, and sacrificed to the necessities of the family. Such as are allowed to hatch, instead of being fed and protected as they would be if honey was abundant, are allowed, while yet weak from the effects of hunger, to wander from the hive, and fall to the earth by hundreds. These effects attend only a scarcity in the early part of the season. The massacre of July and September is quite different. The drones then have age and strength—an effort is apparently first made by the workers to drive them out without proceeding to extremes; they are harassed sometimes for several days; the workers feigning only to sting, or else they cannot, as I never succeeded in seeing but very few dispatched in that way; yet there is evidence proving beyond doubt that the sting is used. Hundreds will often be collected together in a compact body at the bottom of the hive; this mutual protection affording a few hours' respite from their tormentors, who do not cease to worry them. In a few days they are gone, and it is a hard matter to tell what has become of them, at least the majority. If the hive in September is well supplied with honey, a portion of the drones have a longer lease of life given them; I have seen them as late as December. In some seasons, when the best hives are poorly supplied with stores, the ensuing spring the bees will rear no drones, until the flowers yield a good supply. I have known one or two years in which no drones appeared before the last of June; at other times, thousands are matured by the first of May.

OLD QUEEN LEAVES WITH THE FIRST SWARM.

The old queen leaves with the first swarm; as soon as cells are ready in the new hive she will deposit her eggs in them, at first for workers; the number perfected will correspond with the supply of honey and size of the swarm. When the supply fails before leaving the old stock, she remains *there*, and continues laying throughout the season; but the bees matured after the 20th of July (in this section) are not more than sufficient to keep the number good. As many die, or are lost during their excursions, as the young ones will replace; in fact, they often lose rather than gain; so that by the next spring, a hive that has cast no swarm, is no better for a stock than one from which a swarm has issued. We are apt to be deceived by bees clustering outside, towards the latter end of the season, and suppose it hardly possible for them all to get in, when it may be caused by hot weather, full stores, &c.

A YOUNG QUEEN TAKES THE PLACE OF HER MOTHER IN THE OLD STOCK.

In ordinary circumstances, when a swarm has left a stock, the oldest of the young queens is ready to emerge from her cell in about eight or nine days; if no second swarm is sent out, she will take her mother's place, and begin to lay eggs in about ten days, or a little less. Two or three weeks is the only time throughout the whole season, but what eggs can be found in all prosperous hives. Whenever a copious yield of honey occurs, drones are reared; as it becomes scarce, they are destroyed.

The relative number of drones and workers that exist when they are most numerous, doubtless depends on the size of the hive, whether one in ten, or one in thirty.

When a swarm is first hived, the first cells are the size for working; if the hive be very small, and bees numerous, it may be filled before they are fully aware of it, and but few drone-cells constructed; consequently, but few can be raised; whereas if the hive be large, long before it is full, considerable honey will be stored. Cells for storing honey are usually the size for drones; these will be made as soon as the requisite number for workers is provided. An abundant yield of honey during the process of filling a large hive, would therefore cause a great proportion of these cells to be built—the amount of drone-brood being governed by the same cause, is a strong argument against large hives, as affording room for too many of these cells, where an unnecessary number of drones will be reared, causing a useless expenditure of honey, &c.

OTHER THEORIES.

Theories differing materially from the foregoing, are advanced by nearly all writers. One says, "In spring the queen lays about 2,000 eggs of males, resumes it again in August, but during the rest of the intervals she exclusively lays worker eggs. The queen must be at least eleven months old before she begins to lay the eggs of males." Mr. Townley makes the same assertion. Dr. Bevan says, "the great laying of drone eggs usually commences about the end of April." Another author repeats about the same, and appears to have investigated farther, as he has found out that the eggs for the two kinds of bees are germinated separately, and the queen knows when each kind is ready, as well as the workers, &c. Now, I beg leave to differ a little from these authors. Either there exists no difference in the eggs germinated, and any, or all will produce drones or workers, just as they happen to be deposited and fed; or else the periods of laying drone eggs are much more frequent than any writer with which I am acquainted has been willing to allow.

SUBJECT NOT UNDERSTOOD.

I am not anxious to establish a new theory, but to get at facts. If we pretend to understand natural history, it is important that we have it correct; and if we do not understand it, say so, and leave it open for further investigation. It is my opinion that we *know* but very little about this point. I wish to induce closer observation, and would recommend no *positive* decision, until all the facts that will apply have been examined. Whether these drone-egg theories have been too hastily adopted, the reader can decide; I shall offer a few more facts, somewhat difficult to reconcile with them.

First, in relation to the queen being "eleven months old" before laying drone eggs. We *all* agree, I believe, that the old queen goes with the first swarm, and a young one remains in the old stock. Now suppose the first swarm leaves in June, and the old stock yet contains a numerous family. The flowers of buckwheat in August yield a bountiful harvest of honey. This old stock rears a large brood of drones. Is it not proved in this case that the queen was but two months old, instead of eleven? We further agree that young queens accompany second or after-swarms. When these happen to be large and prosperous, they never fail to rear a brood of drones at this season. What is the age of these? I apprehend that this eleven months theory originated in sections where there are no crops of buckwheat raised, or in small quantities. Clover generally fails in August, and May, or June, of another year comes round, before there is a sufficient yield to produce the brood. With these observations *only*, how very rational to conclude that it must be a law of their nature, instead of being governed by the yield of honey, and size of the family? If the periods of drone egg laying are limited to only two or three, it would seem that all queens ought to be ready with this kind of egg, about the same period of the season, but how are the facts?

I would like to inquire what becomes of the first series of drone eggs, the last of April, or the first of May, when the stocks are poorly supplied with honey, or when a family is small and but little honey through the summer? No drone brood is matured in these cases. It is not pretended that the queen has any control over the germination of these eggs, yet somehow she has them ready whenever the situation of the hive will warrant it. Two stocks may have an equal number of bees the first of May; one may have forty pounds of honey, the other four pounds; the latter cannot afford to rear a drone, while the other will have hundreds. Let two stocks have but four pounds each at any time in summer when honey is scarce, now feed one of them plentifully, and a brood of drones is sure to appear, while the other will not produce one.

Whenever stocks are well stored with honey, and full of bees, the first of May will find drone-cells containing brood. If the flowers continue to yield a full supply, these cells may be examined every week from that period till the first swarm leaves, and I will engage that drone brood may be found in all stages from the egg to maturity; and the worker brood the same. In twenty-four days after the first swarm leaves, the last drone eggs left by the old queen will be just about matured. When transferring bees from old to new hives, I generally do it about twenty-one or twenty-two days after the first swarm, (this is the time to avoid destroying the worker-brood; the particulars will be given in another place.) I have transferred a great many, and *never failed* to find a few drones about ready to leave the combs. Whether the swarm had left the last of May, or middle of July, there was no difference, they were on hand.

A very early swarm in good seasons, will often fill the hive, and send out an issue in from four to six weeks: the usual amount of drone-brood may be found in these cases. The following circumstance would appear to indicate that all the eggs are alike, and if they are laid in drone-cells, the bees give the proper food and make drones; if in worker-cells, workers, just as they make a queen from a worker-egg, when put in a royal cell.

In a glass hive, one sheet of comb next the glass, and parallel with it, was full size; about three-quarters of this sheet was worker-cells, the remainder drone-cells. The family had been rather small, but now had increased to a full swarm; a few drones had matured in the middle of the hive. It was about the middle of June, 1850, when I discovered the bees on this outside sheet, preparing it, as I thought, for brood, by cutting off the cells to the proper length. They had been used for storing honey, and were much too long, being about an inch and a half deep. In a day or two after I saw a few eggs in both worker and drone-cells; four or five days afterwards, on opening the door, I found her "majesty" engaged in depositing eggs in the drone cells. Nearly every one already contained an egg; most of these she examined, but did not use them; six or eight, it appeared, were all that were unoccupied; in each of these she immediately deposited an egg. She continued to search for more empty cells, and in doing so, she got on the part of the comb containing worker-cells, where she found a dozen or more empty, in each of which, she laid one. The whole time perhaps thirty minutes. Query? Was her series of drone eggs exhausted just at this time? If so, it would appear that she was not aware of it, because she examined several drone-cells after laying the last one there, before leaving that part of the comb, and acted exactly as if she would have used them had they not been pre-occupied. Did the worker-cells receive some eggs that would have produced drones, but for the circumstance of being deposited in worker-cells? I know we are told that an egg may be transferred from a worker-cell to one for drones, or an egg taken from a drone-cell and deposited in a worker-cell; that the exchange will make no difference, the bee will be just what the first deposit would have made it. How the knowledge for this assertion was obtained, we are not informed, at least of the practical part. That an egg was ever detached from the bottom of one cell safely and successfully deposited in another, without breaking or injuring it in some manner, to make the bees refuse it, permit me at present to doubt.

NECESSITY FOR FURTHER OBSERVATION.

Cannot some experiments, practicable to all, be instituted that will throw more light on this subject? The old hypothesis of limiting drone-egg laying to two or three periods, is evidently at fault.

TWO SIDES OF THE QUESTION.

If we suppose that the eggs are all alike, and the subsequent treatment makes either workers, drones, or queens, and look to analogy for support, we shall find much against, as well as for it. For instance, we find in almost every department of animated nature, that the sex of the germ of a future being is decided before being separated from the parent, as the eggs of fowls, &c. Another fact, some queens (averaging one in sixty or eighty) deposit eggs that produce only drones,[8] whether in worker or drone-cells, proving that sex is decided in this case beyond controversy. Hence it would appear reasonable, if sex was decided by the ovaries of the queen, in one case, it would be in another.

To allow the bees the power of making three kinds of bees from one kind of eggs, which would be virtually constituting a third sex, an anomaly not often found. The drones being males, and workers imperfect females with generative organs undeveloped, renders the anomaly of the third sex unnecessary. On the other side it might be said in reply: That if food and treatment would create or produce organs of generation in the female, by making an egg destined for a worker into a queen, (a fact which all apiarians admit,) why not food and treatment make the drone? Is the difficulty of developing *one* kind of sexual organs greater than another?

Respecting the anomaly of the eggs of some queens producing only drones, the question might be asked, Is this more of an anomaly than that of ordinary queens which are said to germinate eggs in distinct series? It is all out of the usual line. Other animals or insects usually produce the sexes promiscuously. As we are ignorant of causes deciding sex in any case, we must acknowledge mystery to belong to both sides of the question here. The stumbling-block of more than two sexes, which seems so necessary to make plain, is no greater here than with some species of ants, that have, as we are told, king, queen, soldier and laborer. Four distinct and differently formed bodies, all belonging to one nest, and descended from one mother. Whether there are four distinct kinds of eggs producing them, or the power is given to the workers to develop such as are wanted, from one kind, we cannot say. If we make two kinds of eggs, it helps the matter but very little. There is still an anomaly. There is but one perfect female in a nest to germinate eggs, and the myriads produced (being over 80,000 in twenty-four hours, according to some historians) shows that the fecundity of our queen-bee is not a parallel case by any means. And yet they are similar, by having their offspring provided for without an effort of their own.

I shall leave this matter for the present, hoping that *something conclusive* may occur in the course of my experiments, or those of others. At present I am inclined to think that the eggs are all alike, but am not fully satisfied.

I am aware that this matter is of but little value or interest to many, but myself and a few others have "Yankee inquisitiveness" pretty well developed, and would like to *know* how it *was* managed.

As for workers proving occasionally fertile, I have but little to say. After years of close observation directed to this point, I have been unable to discover anything to establish this opinion. Neither have I found the black bees described by some authors. It is true that in the middle or latter part of summer a portion will be much darker than others, and perhaps rather smaller, and some of them with their wings somewhat worn, probably the result of continued labor, peculiar food, or some incidental circumstance.

I have a few times found a humble-bee under the hive, that had entered, and not finding his way out readily, was speedily shorn of his beautiful "locks," and consequently his strength—that is, every particle of hair, down, feathers, bristles, or whatever he had been covered with, was completely removed by the bees, who had no regard for his beautiful alternating stripes of yellow and brown; which left him the very picture of darkness.

CHAPTER IV.

BEE PASTURAGE.

In some seasons the earth is covered with snow much later than others. When this occurs, a greater number of warm days are necessary to melt it, and start the flowers, than otherwise.

SUBSTITUTE FOR POLLEN.

During these warm days, while waiting for the flowers, the bees are anxious to do something. It is then interesting to watch them, and see what will be used as substitutes for pollen and honey. At such times, I have seen hundreds engaged on a heap of sawdust, gathering the minute particles into little pellets on their legs, seeming quite pleased with the acquisition. Rotten wood, when crumbled into powder, and dry, is also collected. Flour, when scattered near the hive, I have known to be taken up in considerable quantities. Some apiarians have fed it to their bees at this season, and consider it a great advantage; I have not tested it sufficient to give an opinion. A substitute for honey is sap from a few kinds of trees, yet it all amounts to but very little. All these unnatural sources are abandoned when the flowers appear.

MANNER OF PACKING IT.

The particular manner of obtaining pollen has been witnessed by but very few persons, as it is generally brushed from their bodies and packed on their legs, while on the wing, thereby preventing a fair chance to inspect operations. When collecting only pollen they alight on the flowers, passing rapidly over the stamens, detaching a portion of the dust, which lodges on most parts of them, to be brushed together and packed into pellets when again on the wing. Thus they keep alternately flying and alighting until a load is obtained, when they immediately return to the hive; each bee bringing several loads in a day. Honey, as it is collected, is deposited in the abdomen, and kept out of sight till stored in the hive.

ALDER YIELDS THE FIRST.

The first material gathered from flowers is pollen. Candle-alder (*Alnus Rubra*)[9] yields the first supply. The time of flowering varies from the 10th of March to the 20th of April. The amount afforded is also variable. Cold, freezing weather frequently destroys a great portion of these flowers after they are out. These staminate flowers are nearly perfected the season previous, and a few warm days in spring will bring them out, even before any leaves appear. When the weather continues fine, great quantities of farina are secured.

The time that bees commence their labors does not govern the time of swarming by any means; this matter depends on the weather through April and May. These remarks apply particularly to this section, Green County, New York, in latitude about 42 degrees. In other places many different trees, shrubs, and herbs, may be found yielding honey and pollen that scarcely exist here, producing far different results.

Our swamps produce several varieties of willow, (salix,) that put out their blossoms very irregularly. Some of these bushes are a month earlier than others, and some of the buds on the same bush are a week or two later than the rest. These also afford only pollen, but are much more dependence than alder, as a turn of cold weather cannot at any time destroy more than a small part. Next comes the aspen, (*Populus Tremuloides*); of this we have more than is necessary for any purpose. It is not a particular favorite with the bees, as but few, comparatively, visit it. It is followed very soon by an abundance of the red maple (*Acer Rubrum*), that suits them better, but this, like the others, is often lost by freezing. The first honey obtained of any account is from the golden willow (*Salix Vitellina*); it yields no pollen, and is seldom injured by frost. Gooseberries, currants, cherries, pear and peach trees, add a share of both honey and pollen. Sugar maple (*Acer Saccharinum*) now throws out its ten thousand silken tassels, beautiful as gold. Strawberries modestly open their petals in invitation, but, like "obscure virtues," are often neglected for the more conspicuous Dandelion, and the showy appearance and flagrant blossoms of the apple-trees, which now open their stores, offering to their acceptance a real harvest.

FRUIT FLOWERS IMPORTANT IN GOOD WEATHER.

In good weather, sometimes a gain of twenty lbs. is added to their stores, during this period of apple-tree blossoms. But we are seldom fortunate enough to have good weather all through this period, it being rainy, cloudy, cool, or windy, which is very detrimental. Sometimes a frost at this time destroys all, and the gain of our bees is reversed, that is, they are lighter at the end than at the beginning of these flowers. Yet this is the season that decides their prosperity for the summer, whether they do *first rate* or otherwise. If good weather now, we expect our first swarms about the first of June; if not, no subsequent yield of honey will make up for this deficiency. We now have a time of several days, from ten to fourteen, in which but few flowers exist. If our hives are poorly supplied when this scarcity occurs, it will so disarrange their plans for swarming, that no preparations are again made much before July, and sometimes not at all. In sections where the wild cherry (*Cerasus Seratina*) abounds, the flowers of this will appear and fill this time of scarcity, which this section annually presents.

RED RASPBERRY A FAVORITE.

The red raspberry (*Rubus Strigosus*) next presents the stamens as the most conspicuous part of the flower, soliciting the embrace of the bee, by pouring out bounteous libations more prized by our industrious insect than wine. For several weeks they are allowed to partake of this exquisite beverage; it is secreted at all hours and in all kinds of weather. When the morning is warm we often hear their cheerful humming among the leaves and flowers of this shrub, ere the sun appears above the horizon. The gentle shower, sufficient to induce man to seek a shelter, is often unheeded by the bee when luxuriating among these flowers; even white clover, important as it is in furnishing the greatest part of their stores, at this season, would be neglected if there was only a full supply of this. Clover begins to blossom with the raspberry, and continues longer. We have an insufficient supply (in this section) in most seasons. Red clover probably secretes as much honey as the white, but the tube of the corolla being longer, the bee appears to be unable to reach it. Yet I have seen a few at work even here but it appeared like slow business. Sorrel, (*Rumex Acetosella*) the pest of many farmers, is brought under contribution, and furnishes the precious dust in any quantity. Morning is the only part of the day appropriated to its collection.

CATNIP, MOTHER-WORT, AND HOARHOUND ARE SOUGHT AFTER.

Catnip, (*Nepeta Cataria*,) Mother-wort, (*Leonurus Cardiaca*,) and Hoarhound, (*Marrubium Vulgare*,) about the middle of June, put forth their flowers, rich in sweetness, and like the Raspberry, the bees visit them at all hours and in nearly all kinds of weather. They last from four to six weeks; the catnip I have known to last twelve in a few instances, yielding honey during the whole time. Ox-eye daisy, (*Leucanthemum Vulgare*,) that beautiful and splendid flower, in pasture and meadow, and worth but little in either, also contains some honey. The flower is compound, and each little floret contains particles so minute, that the task of obtaining a load is very tedious. It is only visited when the more copious honey-yielding flowers are scarce. Snap-dragon,(*Linaria Vulgaris*,) with its nauseous and sickening odor, troubling the farmer with its vile presence, is made to bestow the only good thing about it, except its beauty, upon our insect. The flower is large and tubular, and the bee to reach the honey must enter it; to see the bee almost disappear within the folds of the corolla, one would think that it was about being swallowed, when the hideous mouth was gaping to receive it; but unharmed, soon it emerges from the yellow prison, covered with dust; this is not brushed into pellets on its legs, like the pollen from some other flowers, but a part adheres to its back between the wings, which it is apparently unable to remove, as it remains there sometimes for months, making a cluster outside the hive, appear quite speckled. Bush honey-suckle (*Diervilla Trifida*) is another particular favorite.

SINGULAR FATALITY ATTENDANT ON SILKWEED.

Silkweed (*Asclepias Cornuti*) is also another honey-yielding perennial, but a singular fatality attends many bees while gathering it, that I never yet saw noticed. I had observed during the period this plant was in bloom, that a number of the bees belonging to swarms, before the hive was full, were unable to ascend the sides to the comb; there would be sometimes thirty or more at the bottom in the morning. On searching for the cause, I found from one to ten thin yellow scales, attached to their feet, triangular, or somewhat wedge shape, in size about the twentieth part of an inch. On the longest point or angle, was a black thread-like point, from a sixteenth to an eighth of an inch in length; on this stem was either hooks, barbs, or a glutinous matter, that firmly adhered to each foot or claw of the bee, rendering it useless as far as climbing the sides of the hive was concerned. I found also among bees clustered outside of full hives, this ornament attached, but to them it appeared no inconvenience. Among the scales of wax and waste matter that accumulates about the swarms to the amount of a handful, I found a great many of these scales, which the bees had worked from their feet. The question then arose, were these scales a foreign substance, accidentally entangled in their claws, or was it something formed there by nature, or *rather* an unnatural appendage? It was soon decided. From the number of bees carrying it, I was satisfied that if it was the product of any flower, it belonged to a species somewhat abundant. I set about a close examination of all such as were then in bloom. I found the flowers of the Silkweed, (or Milkweed, as some call it,) sometimes holding a dead bee by the foot, secured by this appendage. Both sepals and petals of this flower are re-curved, that is, turned backward towards the stem, forming five acute angles, or notches, just the thing for a trap for a bee with *strings* of *beads* on its toes; when at work they are very liable to slip a foot into one of these notches; the flower being thick and firm, holds it fast; pulling only draws it deeper into the wedge-like cavity. The bee must either perish or break loose; their instincts fail them in this emergency; they know nothing about getting it out by a gentle pull the other way. I never saw one do it except by accident. By examining the buds of this plant just before opening, I found this fatal appendage, by which great numbers of our bees are lost.[10] When I point out a loss among our bees, I would like to give a remedy; but here I am at a loss, unless all these plants are destroyed, and this is impracticable in many places. After all I am not sure but honey enough is obtained by such bees as do escape, to counterbalance what we lose. This would depend on the amount of honey yielded by other flowers at the same time.

Whitewood (*Liriodendron Tulipifera*) yields something eagerly sought for by the bees, but whether honey, or pollen, or both, I have never been able to ascertain. All the flowers of this kind, with us, are too high. It is very scarce, as well as Basswood, (*Tilia Americana,*)—that in some places is abundant, and yields honey clear and transparent as water, superior in appearance, but inferior in flavor to clover; it also appears much thinner when first collected.

LARGE YIELD FROM BASSWOOD.

During the time this tree is in bloom, a period of two or three weeks in many sections, astonishing quantities are obtained. A person once assured me that he had known "ten pounds collected by one swarm in a day, by weighing the hive in the morning and again at evening." I have some doubt of the statement, and think half the amount would be a good day's work; but I had but a small chance to know, as only a few trees, as a specimen, grow in this section. I have weighed hives during seasons of apple-tree blossoms and buckwheat, the two best yields of honey we have, and three and a half pounds was the best for one day that I ever had. Sumach, (*Rhus Glabra,*) in some sections, affords considerable honey. Mustard (*Sinapis Nigra*) is also a great favorite.

I have now mentioned most of the honey-producing trees and plants that come on before the middle of July. The course of these flowers is termed the first yield. In sections where there are no crops of buckwheat, it constitutes the only full one. Other flowers continue to bloom till cold weather. Where white clover is abundant and the fields are used for pasture, it will continue to throw out fresh flowers, sometimes, throughout the summer; yet the bees consume about all they collect in rearing their brood, &c. Thus it appears in some sections six or eight weeks is about all the time they have to provide for winter.

GARDEN FLOWERS UNIMPORTANT.

In passing along I have not mentioned garden flowers, because the amount obtained here is a small item, compared to the forest and fields—especially ornamental flowers. It is true that the Hollyhock, (*Altha Rosea,*) Mallows, (*Malva Rotundifolia*) and many others yield honey, but what does it amount to? A person expecting his hives to be filled from such a source would very likely be disappointed, especially when many are kept together.

HONEY-DEW.

Honey-dew is said to be a source from whence large collections are made in some places. When or where it appears or disappears is more than I can tell. I have seen the accounts of it, but accounts I have learned to doubt until I find something corroborative in my own experience. I find too many errors copied merely because they happen to be in company with several truths.

Huber discovered many important truths, and has given them to the world; too many writers take it for granted when two points of his are true, the third *must be also*. It is no proof that there is no such article merely because I never discovered it. In the many fruitless endeavors that I have made to get a view of this substance, it may be I have lacked close observation; or possibly there is none showered upon this region; or I may have failed to bring my imagination to assist me to convert common dew into the real article.

SINGULAR SECRETION.

I once discovered bees collecting a secretion unconnected with flowers; but was not honey-dew, as it has been described. I was passing a bush of Witch-hazel, (*Hamamelis Virginiana*,) and was arrested by an unusual humming of bees. At first I supposed that a swarm was about me, yet it was late in the season, (it being about the 25th July.) On close inspection, I found the bush contained numerous warty excrescences, the size and shape of a hickory-nut. These proved to be only a shell—the inside lined with thousands of minute insects, a species of aphis. These appeared to be engaged sucking the juices, and discharging a clear, transparent fluid. Near the stem was an orifice about an eighth of an inch in diameter, out of which this liquid would gradually exude. So eager were the bees for this secretion, that several would crowd around one orifice at a time, each endeavoring to thrust the other away. This occurred several years ago, and I never have been able to find anything like it since; neither have I learned whether it is common in other sections.

SECRETIONS OF THE APHIS.

The liquid ejected by the aphis, (plant louse,) when feeding or sucking the juices of tender leaves, and received by the ants that are always in attendance, is something like it; but in this case the bees were in attendance instead of ants.

This mode of elaborating honey, although not generally collected by bees, perhaps may not be too much out of place here. Also, it may furnish a clue to the cause or substantiate some theory of honey-dew.

These insects (*Aphis*) have been very appropriately termed "ants' cows," as they are regarded by them with the most tender care and solicitude. In July or August, when the majority of the leaves of our apple trees are matured, there is often a few sprouts or suckers about the bottom or trunk, that continue growing and putting out fresh leaves. On the under side of these, you will find the *aphis* by hundreds, of all sizes, from those just hatched to the perfect insect with wings. All appear to be engaged in sucking the bitter juice from the tender leaf and stalk. The ants are among them by scores. (They are often accused by the careless observer of the injury, instead of the *aphis*.) Occasionally there will issue from their abdomen a small, transparent globule, which the ant is ever ready to receive. When a load is obtained it descends to the nest; others may be seen going and returning continually. Many other kinds of trees, shrubs and plants are used by the ants as "cow pasture," and most kinds of ants are engaged in this dairy business.[11] Would the bees attend on the *aphis* for this secretion, (for it appears to be honey,) if the ant was not there first? Or if there were no ants or bees, would this secretion be discharged, and falling on the leaves below them, be honey-dew? If they were situated on some lofty trees, and it lodged on the leaves of small bushes near the earth, it would, with some authors.

These questions I shall not answer, at present. As for theory, I shall probably have enough before I get through, where I hope the subject may be more interesting.[12]

We will now return to the flowers, and see what few there are yet to appear, after the middle of July. The button-ball bush (*Cephalanthus Occidentalis*) is now much frequented for honey. Also, our vines, melons, cucumbers, squashes, and pumpkins. The latter are visited only in the morning, and honey is the only thing obtained; notwithstanding the bee is covered with farina, it is not kneaded into pellets on its legs. I have seen it stated that bees never get honey early in the morning, but pollen instead. Now it is not best always to take our word, who pretend to know all about it, but look for yourselves into some of these matters. Take a look some warm morning, when the pumpkins are in bloom, and see whether it is honey or pollen they are in quest of. Also please make an observation when they are at work on the red raspberry, motherwort, or catnip; you will thus ascertain a fact so easily, that you will wonder any one with the least pretension to apiarian science could be ignorant of it. I mention this, not because it is of much importance in itself, but to show the fallibility of us all, as we sometimes copy the mistaken assertions of others.

ADVANTAGES OF BUCKWHEAT.

Under some circumstances, clover will continue to bloom through this part of the season; also, a few other flowers; but I find by weighing, a loss from one to six pounds, between the 20th July and the 10th of August, when the flowers of buckwheat begin to yield honey, which generally proves a second harvest. In many places it is their main dependence for surplus honey. It is considered by many an inferior quality. The color, when separated from comb, resembles molasses of medium shade. The taste is more pungent than clover honey; it is particularly prized on that account by some, and disliked by others for the same reason. In the same temperature it is a little thicker than other honey, and is sooner candied.

AMOUNT OF HONEY COLLECTED FROM IT.

Swarms issuing as late as the 15th July, when they commence on buckwheat, sometimes contain not over five pounds of stores, and yet make good stocks for winter, whereas, without this yield, they might not live through October. It fails about once in ten years. I have known a swarm to gain in one week sixteen pounds, and construct comb to store it at the same time. At another time I had a swarm issue the 18th August, that obtained thirty pounds in about eighteen days. But such buckwheat swarms, in ordinary seasons, seldom get over fifteen pounds. The flowers last from three to five weeks. The time of sowing the grain varies in different sections, from the 10th of June to the 20th July. Farmers wish to give it just time to ripen before frost, as the yield of grain is considered better, but as the time of frost is a matter of guess-work, some will sow several days earlier than others. Whenever an abundant crop of this grain is realized, a proportionate quantity of honey is obtained.

DO BEES INJURE THE CROP?

Many people contend that bees are an injury to this crop, by taking away the substance that would be formed into grain. The best reasons for this opinion that I have obtained are these: "I believe it, and have thought so a long time." "It is reasonable if a portion of this plant is taken away by the bees, there must be a less quantity of material left for the formation of seed, &c." Most of us have learned that a person's opinion is not the strongest kind of proof, unless he can exhibit substantial reasons for it. Are the above reasons satisfactory? How are the facts? The flowers expand, and a set of vessels pour into the cup or nectary a minute portion of honey. I am not aware that any one contends that the plant has another set of vessels prepared to again absorb this honey and convert it into grain. But strong testimony proves very plainly that it never again enters the stalk or flower, but evaporates like water. We all know that animal matter when putrid will be dissolved into particles small enough to float in the atmosphere, too minute for the naked eye. When passing off in this way this real flesh and blood would escape notice perhaps altogether, and never be detected, were it not for the olfactories, which on some occasions notify us of its presence very forcibly. In passing a field of buckwheat in bloom, by the same means we are assured of the presence of honey in the air. Now what is the difference whether this honey passes off in the air, or is collected by the bees? If any difference, the advantage appears to be in favor of the bees getting it, for the reason that it thus answers another important end in the economy of nature, consistent with her provisions in ten thousand different ways of adapting means to ends. Most breeders of domestic animals are aware of the deteriorating qualities induced by in-and-in breeding; a change of breed is found necessary for perfection, &c.

ARE NOT BEES AN ADVANTAGE TO VEGETATION?

Vegetable physiology seems to indicate a similar necessity in that department. The stamens and pistils of flowers answer the different organs of the two sexes in animals. The pistil is connected with the ovaries, the stamens furnish the pollen that must come in contact with the pistil; in other words, it *must be impregnated* by this dust from the stamens, or no fruit will be produced. Now if it be necessary to change the breed, or essential that the pollen produced by the stamens of one flower shall fertilize the pistil of another, to prevent barrenness, what should we contrive better than the arrangement already made by Him who knew the necessity and planned it accordingly? And it works so admirably, that we can hardly avoid the conclusion *that bees were intended for this important purpose*! It is thus planned! Their wants and their food shall consist of honey and pollen; each flower secretes but little, just enough to attract the bee; nothing like a full load is obtained from one; were it thus, the end in view would not be answered; but a hundred or more flowers are often visited in one excursion; the pollen obtained from the first may fertilize many, previous to the bees' returning to the hive; thus a field of buckwheat may be kept in health and vigor in its future productions. A field of wheat produces long slender stalks that yield to the influence of the breeze, and one ear is made to bestow its pollen on a neighboring ear several feet distant, thereby effecting just what bees do for buckwheat. Corn, from its manner of growth, the upright stalk bearing the stamens some feet above the pistils, on the ears below, seems to need no agency of bees; the superabundant pollen from the tassel is wafted by the winds rods from the producing stalk, and there does its office of fertilizing a distant ear, as is proved by different varieties mixing at some distance. But how is it with our vines trailing on the earth, a part of these flowers producing stamens, the other only pistils? Now it *is absolutely essential* that pollen from the staminate flowers shall be introduced into the pistillate to produce fruit; because if a failure occurs in this matter the germ will wither and die. Here we have the agent ready for our purpose; these flowers are visited by the bee promiscuously; no pollen (as was said) is kneaded into pellets, (particularly that from pumpkins,) but it adheres to every part of their body, rendering it next to impossible for a bee thus covered with dust to enter the pistillated flower without fulfilling the important duty designed, and leave a portion of the fertilizing dust in its proper place. Hence it is reasonably inferred by many, that if it was not for this agent among our vines, the uncertainty of a crop from non-fertilization would render the cultivation of them a useless task.

When the aphis is located on the stalk or leaf of a plant it is furnished with means to pierce the surface and extract the juices essential to the formation of the plant, thereby preventing vigorous growth and a full development.

This idea is too apt to be associated with the bee when she visits the flower, as if she was armed with a spear, to pierce bark or stem and rob it of its nourishment. Her real structure is lost sight of, or perhaps never known; her slender brush-like tongue folded closely under her neck, and seldom seen except when in use, is not fitted to pierce the most delicate substance; all that it can be used for is to sweep or lick up the nectar as it exudes from the pores of the flower, secreted, it would seem, for no other purpose but to attract her—while there she obtains nothing but what nature has provided for her and given her the means of obtaining, and the most delicate petal receives no injury.

During an excursion the bee seldom visits more than a single species of flower; were it otherwise, and all kinds of flowers were visited promiscuously, by fertilizing one species with the pollen from another, the vegetable kingdom would be very likely to get into confusion. Writers, when noticing the peculiarity of instinct governing the bee here, cannot be content always, but must add other marvels. They follow this trait into the hive, and make her store every kind by itself there. Relative to honey it is not an easy matter to be positive; but pollen is of a variety of colors, generally yellow, yet sometimes pale-green, and reddish or dark-brown. Now I think a little patient inspection would have satisfied any one that two kinds *are* sometimes packed in one cell, and prevented the assertion to the contrary. I will admit that two colors are seldom found packed together, but sometimes will be. I have thus found it, and it has entirely ruined that theory for me.

A TEST FOR THE PRESENCE OF QUEEN DOUBTED.

It is further asserted that if a hive loses its queen "no pollen is collected." Also, "that such quantities are sometimes collected, and fill so many cells, that too little room is left for brood, and the stock rapidly dwindles away in consequence." The first of these assertions has been given as a test to decide whether the hive contains a queen or not. Now my bees have such a habit of doing things wrong that the above is no test whatever. It is made to appear very well in theory, but wants the truth in practice. I will say what I have known on this point, and perhaps clear up the difficulty of a stock containing an unusual quantity of bee-bread with the honey, and instead of being the cause of its having but few bees, it is the effect. Stocks and sometimes swarms lose their queen in the swarming season, (the particulars will be given in another place,) when, instead of remaining idle, the usual quantity of both *pollen and honey is collected* (unless the family is very small). There being no larvæ to consume the bread, the consequence is, more than half the breeding cells will contain it; they will be packed about two-thirds full, and finished out with honey. I have known a large family left under such circumstances, and about all the cells in the hive would be occupied. Whereas, in a stock containing a queen and rearing brood, *a portion of the combs will be used for this purpose until the flowers fail*, and then such comb will be found empty.

AN EXTRA QUANTITY OF POLLEN NOT ALWAYS DETRIMENTAL.

To test whether this extra quantity of bee-bread was so *very* detrimental, I have introduced into such hive in the fall a family with a queen and wintered them in it, and watched their prosperity another year, and never found them less profitable on that account. I am so well satisfied of this, that whenever I now have a hive in such a situation, it is a rule to introduce a swarm.

It is calculated, I believe, generally, that when medium-sized hives are full, about seven-eighths of the cells are made the proper diameter for raising the workers, the remainder for drones, except a few for queens. Here is one circumstance I do not remember to have seen mentioned, and that is, bee-bread is generally packed exclusively in the worker cells. I would say always; but I would do better to be careful, especially as I find my bees doing things so differently from some others. I might as well remark here, that when taking combs from a hive filled with honey, if such pieces were selected as contained only the large or drone cells, but little risk of bee-bread would occur; of the other combs, the outside sheets and the corners of the others near the top are the next best. The sheets of comb used principally for raising workers, and the cells next those so used, for an inch or two in width, are nearly all packed with pollen, and much of it will remain, when the breeding season is past. Smaller portions are found in the worker cells in nearly all parts of the hive; even the boxes will sometimes contain a little.

MANNER OF PACKING STORES.

In a glass hive, the bees may be seen depositing their load of pollen; the legs holding the pellets are thrust into the cell, (not their heads), and a motion like rubbing them together is made for a half minute, when they are withdrawn, and the two little loaves of bread may be seen at the bottom. This bee appears to take no farther care about them, but another will soon come along, and enter the cell head first, and pack it close; this cell is filled about two-thirds of its length in this way, and when sealed over a little honey is used to fill it out.

PHILOSOPHY IN FILLING A CELL WITH HONEY.

To witness the operation of depositing honey, a glass hive or box is requisite; the edges of the combs will be attached to the glass—when honey is abundant, most of these cells next the glass will contain some. Now is the time to see the operation, glass forming one side of such as are in contact, &c. The bee may be seen to enter the cell till it reaches the bottom; with its tongue, the first particle is deposited, and brushed into the corners or angles, carefully excluding all the air from behind it—as it is filled, that next the sides of the cell is kept in advance of the centre. The bee does not put its tongue in the centre and pour out its load there, but carefully brushes the sides as it fills, excluding every particle of air, and keeps the surface concave instead of convex. This is just as a philosopher would say it should be. If it was filled at once and no care taken to attach it to the sides, why, the external air would never keep it there, which it does effectually when of ordinary length. When the cell is about one-fourth of an inch deep they often commence filling it, and as it is lengthened they add to it, keeping it within an eighth of an inch of the end; it is never quite full till nearly sealed over, and often not then. In cells of the worker size, the sealing seldom touches the honey. But in the size for drones the case is different; the honey on the end touches the sealing, about half the diameter on the lower side; it is kept in the same shape while being filled; but being somewhat larger, the atmospheric pressure is less effectual in keeping the honey in its place; consequently, when they commence sealing these cells they begin on the lower side and finish at the top.

LONG CELLS SOMETIMES TURNED UPWARD.

When storing honey in boxes, cells of this size are usually much longer, in which case they are crooked, the ends turning upward, sometimes half an inch or more; this, of course, will prevent the honey from running, but if the box is taken off and turned over before such cells are sealed, they are very sure to spill most of their contents. The cells in the breeding apartment, of ordinary length, will hold the honey well enough as long as horizontal; but turn the hive on its side, and bring the open end downward, in hot weather, or break out a piece and hold it in that position, the air will not sustain it in them, but will, in the size suitable for workers.

When the hive is fully supplied with bees and honey, (unless destitute of a queen,) I never examined one, winter or summer, but it had a number of unsealed cells containing honey, as well as pollen; it is so when they have stored fifty pounds in boxes, even when so crowded for room as to store honey outside or under the bottom-board; ever having some cells open for a ready supply.

Young swarms seem unwilling to construct combs faster than needed for use; it would appear, at first thought, to be a lack of economy. When no honey is to be obtained and nothing to do, then it would seem to be a fine chance for getting ready for a yield; but this is not *their* way of doing business; whether they cannot spare the honey already collected to elaborate the wax, or whether they find it more difficult to keep the worms from a large quantity of comb, I shall not decide. Of this I am satisfied, that it is better arranged by their instincts, than we could do it. Large swarms, when first located, if honey is abundant, will extend their combs from top to bottom in a little more than two weeks; but such hive is not yet full; some sheets of comb may contain honey throughout their whole length, and not a cell be sealed over; but, however, they generally find time to finish up within a few inches of the lower end as they proceed. Whenever unfinished cells contain honey, it will generally be removed soon after the flowers fail, and used before that which is sealed; and the cells will remain empty till another year.

IS A DRY OR WET SEASON BEST FOR HONEY?

The inquiry is often made, "What kind of season is best for bees, wet or dry?" This point I have watched very closely, and have found that a medium between the two extremes produces most honey. When farmers begin to express fears of a drought, then is the time (if in the season of flowers) that most honey is obtained; but if dry weather passes these limits, the quantity is greatly diminished. Of the two extremes, perhaps very wet is the worst.

HOW MANY STOCKS SHOULD BE KEPT.

"What number of stocks can there be kept in one place?" is another question often asked. This is like Mr. A. asking farmer B. how many cattle could be pastured in a lot of ten acres. Farmer B. would first wish to know how much pasture said lot would produce, before he could begin to answer; since one lot of that size might produce ten times as much as the other. So with bees, one apiary of two hundred stocks might find honey in abundance for all, and another of forty might almost starve. Like the cattle, it depends on pasture.

THREE PRINCIPAL SOURCES OF HONEY.

There are three principal sources of honey, viz.:—clover, basswood, and buckwheat. But clover is the only universal dependance; as that is almost everywhere, to some extent, in the country. Buckwheat in some places is the main source; in others, basswood, which is of brief duration. Where all three are abundant, there is the true El Dorado of the apiarian! With plenty of clover and buckwheat, it is nearly as well. Even with clover alone, enormous quantities of honey are obtained. I have said what was our dependence in this section. I will further say that within a circle of three or four miles, there are kept about three hundred stocks. I have had for several years, three apiaries about two miles apart, averaging in spring a little more than fifty in each. When a good season for clover occurs, as many more would probably do equally well, but in some other seasons I have had too many; on an average nearly right. When clover furnishes too little honey for the number, buckwheat usually supplies more than is collected. Of surplus honey, the proportion is about fifteen pounds of buckwheat to one of clover. I have now been speaking of large apiaries. There can hardly be a section of country found, that man can procure his living, but what a few stocks would thrive, even if there were no dependence on the sources just mentioned. There will be some honey-yielding flowers in nearly all places. The evil of over-stocking is of short duration, and will work its own cure speedily. Some judgment is required here as well as in other matters.

Another question of some interest, is the distance that a bee will travel in search of honey in flowers—it is evident that it will be farther than they will go to plunder a stock. I have heard of their being found seven miles from home. It was said they ascertained, by sprinkling flour on them as they left the hive in the morning, and then saw the same bees that distance away. When we consider the chances of finding a bee even one mile from the hive thus marked, it appears like a "poor look;" and then pollen the color of flour might deceive us. It is difficult to prove that bees go even two miles. Let us say we guess at it, for the present.

CHAPTER V.

WAX.

The careless, unreflecting observer, when seeing the bees enter the hive with a pellet of pollen on each of their posterior legs, is very apt to conclude that it must be material for comb, as it appears unlike honey. So little regard is paid to the matter by many people, that they are unable to imagine any other use for it. Others suppose that it will change from that to honey, after being stored a time in the hive, and wonder at the curious phenomenon; but when asked how long a time must elapse before it takes place, they cannot tell exactly, but they "have found cells where it began to change, as a portion near the outer end of the cell had become honey, and no doubt the remainder would in time." It has been remarked that cells were only filled about two-thirds full of this, and finished with honey; now when any one finds a cell filled to the brim with pollen, and no honey, such reasoning will apply better. If this was the case, by examining at different periods through the summer, we certainly should find some cells before the change had commenced, instead of their always being in just this stage of transition.

IS POLLEN CONVERTED INTO WAX?

As for pollen being converted into wax or comb, a simple question will show its fallacy. Do not the bees belonging to a hive that is full of combs, and no more wax for that purpose needed, bring home as much and often more pollen than one half full? Any person who has watched two such hives five minutes when busily engaged at work, can answer. It is evident, then, that pollen is for something else besides wax.

HOW IS IT OBTAINED?

The inquiry is now made, "Where do they get it from, if not from pollen?" I might with propriety answer, they don't get it at all. "Stop, there, if you please; if you expect us to credit you, you must not give us too much absurdity." Well, let me ask a question. Do cattle when grazing actually obtain flesh, bone, &c., or only the materials from which these parts are secreted? As to the production of wax, I believe all close observers (that I have found) agree that it is a secretion natural only to the bee. With the ox, fruit, grain, or grass may be converted into tallow; with the bee, honey and syrup made of sugar may be converted into wax. These are probably the only two substances yet discovered from which they extract it. Some writers have pretended that pollen is also used, but they have failed to prove that the old bees consume it at any time; which they must in this case if it is converted into wax. From experiments related by Huber, either of these substances, mixed with a little water, is all sufficient for its production. From experiments of my own, I am satisfied that he is correct. The experiment is tried by shutting up a swarm when first hived; feeding them with honey—a few of the bees will probably have some pollen, though not enough to make a comb three inches square, yet it is something—and to be certain, time must be given them to exhaust it. In three or four days take out the bees and remove the combs; inclose them again, and feed with honey as before. Repeat the process, until satisfied that no pollen is needed in the composition of wax. Huber removed the combs "five times," with the same result at every trial. Whenever bees are *confined* in hot weather, *air and water are absolutely necessary*.

We will now describe the first appearance of wax, and how it is produced. When a swarm of bees is about leaving the parent stock, three-fourths or more of them will fill their sacks with honey. When located in their new home, of course no cells exist to hold it; it must remain in the stomach or sack for several hours. The consequence is, that thin white scales of wax the sixteenth of an inch in diameter, somewhat circular, are formed between the rings of the abdomen, under side. With the claws of one of their hind legs one of these is detached and conveyed to the mouth, and there pinched with their forceps or teeth, until one edge is worked somewhat rough; it is then applied to the comb being constructed, or to the roof of the hive. The first rudiments of comb are often applied within the first half hour after the swarm is hived. In the history of insects before noticed, is a minute account of the first foundation of combs, somewhat amusing, if not instructive.

HUBER'S ACCOUNT OF A COMMENCEMENT OF COMB.

Huber, it is said, "having provided a hive with honey and water, it was resorted to in crowds by bees, who, having satisfied their appetite, returned to the hive. They formed festoons, remained motionless for twenty-four hours, and after a time scales of wax appeared. An adequate supply of wax for the construction of a comb having been elaborated, one of them disengaged itself from the centre of the group, and clearing a space about an inch in diameter, at the top of the hive, applied the pincers of one of its legs to its side, detached a scale of wax, and immediately began to mince it with the tongue. During the operation, this organ was made to assume every variety of shape; sometimes it appeared like a trowel, then flattened like a spatula, and at other times like a pencil, ending in a point. The scale, moistened with a frothy liquid, became glutinous, and was drawn out like a riband. This bee then attached all the wax it could concoct to the vault of the hive, and went its way. A second now succeeded, and did the like; a third followed, but owing to some blunder did not put the wax in the same line with its predecessor; upon which another bee, apparently sensible of the defect, removed the displaced wax, and carrying it to the former heap, deposited it there, exactly in the order and direction pointed out." Now I have some objections to make to this account. First, in the usual course of swarming, it is unnecessary to provide the honey and water, as they come laden with honey from the parent stock. Next, to form festoons and remain motionless twenty-four hours to concoct the wax, is not the way they generally manage affairs. They either swallow the honey before leaving home long enough to have the wax ready, or less time than twenty-four hours is needed to produce it. I have frequently found lumps, half the size of a pin's head, attached to the branch of a tree where they had clustered, when they had not been there over twenty-five minutes. I have had occasion a few times to change the swarm to another tenement, an hour or two after being hived, and found places on the top nearly covered with wax. How it was managed to see a bee quit the "group," is more than I can comprehend; and then the tongue to be the only instrument used to mould the scale of wax, is another difficulty; to witness the whole process minutely in this stage of comb-making has never been my good fortune, and I am sometimes inclined to doubt the success of others. I have had glass hives, and put swarms in them, and always found the first rudiments of comb so entirely covered with bees as to prevent my seeing anything.

BEST TIME TO WITNESS COMB-MAKING.

The only time when I have witnessed the process with any degree of satisfaction is when the combs approach the glass, and but few bees in the way; then, by watching patiently a few minutes, some part of the process may be seen.

MANNER OF WORKING WAX.

Transferring the swarms to different hives from one to forty-eight hours after being hived, will show their progress. I have found that wax is attached to the top of the hive at first promiscuously, that is, without the least order, until some of the blocks or lumps are sufficiently advanced for them to begin cells. The scales of wax are welded on the edge quite thick, without regard to the shape of the cell, then an excavation is made on one side for the bottom of a cell, and two others on the opposite side; the division between them exactly opposite the centre of the first. When this piece is an inch or two in length, two other pieces at equal distances on each side are commenced. If the swarm is large, and honey abundant, it is common for two pieces of comb to be started at one time on different parts of the top; the sheets in the two places are often at right angles, or any other way, just as chance happens to give direction. The little lumps that are placed at random at first are all removed as they advance.

While the combs are in progress, the edges are always kept much the thickest, and the base of the cell is worked down to the proper thickness with their teeth, and polished smooth as glass. The ends of the cell also, as they lengthen them, will always be found much thicker than any other part of it when finished.

When two combs approach each other in the middle of the hive at nearly right angles, an edge of comb is left there; but when an obtuse angle, the edges are generally joined, making a sheet of crooked comb. It is evident where the two combs join, there must be some irregular cells unfit for rearing brood.

CROOKED COMBS A DISADVANTAGE.

These few irregular cells have been considered a great disadvantage. It is thought, or pretended, that there is a vast difference between the prosperity of a stock with straight combs and one with crooked ones. To avoid them, or cause the bees to make them all straight, has given rise to much contrivance, as if a few such cells could effect much. Suppose there were a dozen sheets of comb in a hive, and each one had a row or more of such irregular cells from top to bottom, what proportion would they hold to those that were perfect? Perhaps not one in a thousand. Hence we infer that in a hive of the proper size, the difference in amount of brood never could be perceived. This is the only difference it can make, because such cells can be used for storing honey as well as others. But sometimes there will be corners and spaces not wide enough for two combs, and too wide for one of the proper thickness for breeding. As bees use all their room economically, and generally at the best advantage, a thick comb will be the result. It is said they never use such thick combs for breeding. How are the facts? I have just such a space in a glass hive; one comb two inches thick. How is it managed? Towards fall this sheet is filled with honey; the cells outside are lengthened until there is just room for a bee to pass between them and the glass, when they are sealed over. In spring these long cells are all cut down (except at the top and upper corners) to the proper length for breeding, and used for this purpose. This has been done for five years in succession.

I will grant that there is a little waste room in such spaces, for part of the year. It amounts to but little, as it is only outside. They are necessitated to make such combs, because the inside combs, if built in a breeding apartment, however crooked one may be, the next one will generally match it, the right distance from it. But when they are built expressly for storing honey, in such as are made in boxes, the right distance is not so well preserved; hence it is not recommended to compel bees to use such storing apartment for breeding. But suppose we should compel a swarm to labor under these disadvantages, I should not apprehend such disastrous results, (providing they have a proper proportion of worker cells,) as no swarms, or even no surplus honey, as has been represented. Imagine a hive filled with combs that are all too thick, and room wasted when cut down, to the amount of one-fourth of all that is in the hive. Now here are combs enough left to mature three-fourths as many bees as in an ordinary hive, where all are right. We can now suppose a good swarm will bring home the same amount of honey as though it belonged to other hives; only three-fourths as much can be fed to the brood, and stored in the hive; and the result ought to be, that we get a quarter more surplus honey in boxes. Even if we get no swarm, I cannot see how our surplus honey can be less, as in this case there would be more bees at all times than in a hive that had been reduced by swarming.

Does experience substantiate the theory that stocks with crooked combs are as profitable as when they are straight? When combs are built expressly for breeding, I could never discover any difference. Any person can easily test it by a little observation; not by taking a solitary instance of only one hive, because some other cause might produce the result. Take a half-dozen at least with straight combs, and as many with them crooked; have them all alike in other respects, and carefully watch the result. I think you will have but little interest which way the combs are made, providing *they are made*, as far as profit is concerned. It is true, it would gratify order to have them all straight, and if it was not attended with more trouble than the result would pay for, it would be well to have them so.

In ordinary circumstances, when a swarm is first hived, they set about comb-making immediately; yet sometimes they will remain two days, and not make a particle. I have known them to swarm out and cluster in the usual way, and when rehived, commence at once. This seems to prove that they can retain the wax, or prevent secreting it, till wanted. This seldom occurs.

UNCERTAINTY IN WEIGHT OF BEES.

A large swarm will probably carry with them some five or six pounds of honey from the parent stock. I only guess at this, because I am uncertain what the bees weigh exactly. "I can tell you," some one exclaims, "I saw some weighed,—so many weigh just eight ounces." Are you sure there was nothing but bees weighed? Was there no honey, bee-bread, fæces, or other substance, that might deceive you? "Can't say; I never thought of that!" Now it is important, if we weigh bees to know *their* weight, to be sure we weigh nothing else. It is evident, that if five thousand weigh three pounds, when nothing is in their sacks, they would weigh, when filled with honey, several pounds more. Hence, the fallacy of judging of the size of a swarm by weight, as one swarm might issue with half the honey of another. Perhaps eight pounds, for large swarms, might be an average for bees and honey. This honey, whatever it amounts to, cannot be stored till combs are constructed to hold it. This principle holds good till the hive is full. That is, whenever they have more honey than the combs will hold, if there is room in the hive, they construct more. But they seem to go no farther than this in comb-making. However large the swarm may be, this compulsion appears necessary to fill the hive. Drone-cells are seldom made in the top of the hive, but a part are generally joined on the worker-cells, a little distance from the top; others near the bottom. There seems to be no rule about the number of such cells. Some hives will contain twice the number of others. It may depend on the yield of honey at the time; when very plenty, more drone-cells, &c. If the hive be very large, no doubt an unprofitable number would be constructed. Where the large and small cells join, there will be some cells of irregular shape; some with four or five angles; the distance from one angle to the other is also varied. Even where two combs of cells the same size join, making a straight comb, they are not always perfect.

SOME WAX WASTED.

When constructing comb, they are constantly wasting wax, either accidentally or voluntarily. The next morning after a swarm is located, the scales may be found, and will continue to increase as long as they are working it; the quantity often amounts to a handful or more. It is the best test of comb-making that I can give. Clean off the board and look the next morning, you will find the scales in proportion to their progress. Some will be nearly round as at first; others more or less worked up, and a part will be like fine sawdust.

Huber and some others have divided the working bees into different classes, denominating some wax-workers, others nurses, and pollen gatherers, &c. It may be partially true, but how it was found out is the mystery.

The angles in the cells used for brood, are gradually filled, and after a time become round, both at the ends and sides.

WATER NECESSARY TO COMB-MAKING.

Whenever bees are engaged making comb, a supply of water is absolutely necessary. Some think it requisite in rearing brood. It may be needed for that, or it may be required for both purposes; but yet I have doubts if a particle is given to the young bee, besides what the honey contains. June, and first part of July, and most part of August (the season of buckwheat,) are periods of extensive comb-making; they then use most water; breeding is carried on from March till October, and as extensively in May, perhaps more so, than in August, yet not a tenth part of the water is used in May.

I have known stocks repeatedly to mature brood from the egg to the perfect bee, when shut in a dark room for months, when it was impossible to obtain a drop; also stocks that stand in the cold, (if good,) will mature some brood whether the bees can leave the hive or not. These facts prove that some are reared without water. As they get sufficient honey to require more comb to store it, they will at the same time have a brood; and it is easy to guess they need it for brood as comb, without a little investigation. This much is certain, that they use water at such times for some purpose, and when no pond, brook, spring, or other source is within convenient distance, the apiarian would find it economy to place some within their reach, as it would save much valuable time, if they would otherwise have to go a great distance, when they might be more profitably employed; it always happens in a season of honey. It should be so situated that the bees may obtain it without jeopardizing their lives;—a barrel or pail has sides so steep that a great many will slip off and drown. A trough made very shallow, with a good broad strip around the edge to afford an alighting place, should be provided. The middle should contain a float, or a handful of shavings spread in the water with a few small stones laid on them to prevent their being blown away when the water is out, is very convenient. A tin dish an inch or so in depth, will do very well. The quantity needed may be ascertained by what is used—only give them enough, and change it daily. I have no trouble of this kind, as there is a stream of water within a few rods of the hives; but I have an opportunity to witness something of the number engaged in carrying it. Thousands may be seen (in June and August) filling their sacks, while a continual stream is on the wing, going and returning.

REMARKS.

The exact and uniform size of their cells is perhaps as great a mystery as anything pertaining to them; yet, we find the second wonder before we are done with the first. In building comb, they have no square or compass as a guide; no master mechanic takes the lead, measuring and marking for the workmen; each individual among them is a finished mechanic! No time is lost as an apprentice, no service given in return for instruction! Each is accomplished from birth! All are alike; what one begins, a dozen may help to finish! A specimen of their work shows itself to be from the hands of master workmen, and may be taken as a model of perfection! He, who arranged the universe, was their instructor. Yes, a profound geometrician planned the first cell, and knowing what would be their wants, implanted in the sensorium of the first bee, all things pertaining to their welfare; the impress then given, is yet retained unimpaired! They need no lectures on domestic economy to tell them, by using the base of one set of cells on one side of their combs, for the base of those on the opposite, will save both labor and wax; no mathematician that a pyramidal base, just three angles, with just such an inclination, will be the exact shape needed, and consume much less wax than round or square—that the base of one cell of three angles, would form a part of the base of three other cells on the opposite side of the comb—that each of the six sides of one cell forms one side of six others around it—that these angles and these only would answer their ends.

"The bees appear," says Reaumer, "to have a problem to solve, which would puzzle many a mathematician. A quantity of matter being given, it is required to form out of it cells, which shall be equal, and similar, and of a determinate size, but the largest possible with relation to the quantity of matter employed, while they shall occupy the least possible space!"

How little does the epicure heed, when feasting on the fruits of their industry, that each morsel tasted must destroy the most perfect specimens of workmanship! that in a moment he can demolish what it has taken hours, yes days, perhaps weeks, of assiduous toil and labor, for the bees to accomplish!

CHAPTER VI.

PROPOLIS.

WHAT USED FOR.

This substance is first used to solder up all the cracks, flaws, and irregularities about the hive. A coat is then spread over the inside throughout; when the hive is full, and many bees cluster outside, the latter part of summer, a coat of it is also spread there. An additional coat it seems is annually applied, as old hives will be coated with a thickness proportionate to its age, providing it has been occupied with a strong family. Huber has said it was also used to strengthen the cells when first made, by mixing it with the wax. If it was their practice at that time, the practice has been abandoned by our bees to a great extent. I have made examinations when comb was first made, when it contained eggs, and when it contained larvæ, and have never been able to find anything other than pure wax composing it. After a young bee has matured in a cell, the coating or cocoon that it leaves is of a dark color, somewhat resembling it, and may have given rise to the supposition. How the article is obtained, appears to be the mystery. This is a subject about which apiarians have failed to agree. A few contend that it is an elaborated substance; while others assert it to be a resinous gum, exuding from certain trees, and collected by the bees like pollen. It differs materially from wax, being more tenacious, and when it gets a little age, much harder.

IS IT AN ELABORATE OR NATURAL SUBSTANCE?

No modern observer has ever been able to detect the bees in the act of gathering it.

HUBER'S OPINION.

Huber tells us, that "near the outlet of one of his hives, he placed some of the branches of the poplar, which exuded a transparent juice, the color of garnet. Several workers were soon seen perched upon these branches,— having detached some of this resinous gum, they formed it into pellets, and deposited them in the baskets of their thighs; thus loaded, they flew to the hive, where some of their fellow-laborers instantly came to assist them in detaching this viscid substance from their baskets." Some of our modern apiarians have doubted this account of Huber's. Now, in the absence of anything positive on this subject, I am inclined to adopt this theory; that it is a resin or gum produced by trees. (I cannot say that I am exactly satisfied with the story of bringing the "branches and laying them by the hive," &c.) That bees gather it in its natural state, is in accordance with my own observation.

FURTHER PROOF.

Our first swarms that issue in May, or first of June, seldom use much of the article pure for soldering and plastering; but instead, a composition, the most of which is wax. I have noticed at this season, when old pieces of boards that had been used for hives, were left in the sun, that this old propolis would become soft in the middle of the day. Here I have frequently seen the bees at work, packing it upon their legs; it was detached in small particles, and the process of packing was seen distinctly, as the bee did not fly during the operation, as in the case of packing pollen. It is asserted that when bees need it they always have it, indicating that they can elaborate it like wax. I can see no reason why they do not need it in June as much as August; yet, in the latter month, they use more than a hundred times the quantity. At this time, they manifest no disposition to gather any from the old boards, &c. It would seem they prefer the article new, which they now have in abundance. Boxes filled in June contain but very little, sometimes none. Why not, if they have enough of it? but when filled in August, they always have the corners, and sometimes the top and sides, lined with a good coat. Cracks, large enough for bees to pass through, are sometimes completely filled with it. In this season, a little before sunset of some fair day, I have frequently seen the bees enter the hive with what I supposed to be the pure article on their legs, like pollen, except the surface, which would be smooth and glossy; the color much lighter than when it gets age. I have also seen them through the glass inside, when they seemed unable to dislodge it themselves, like pollen, and were continually running around among those engaged in soldering and plastering; when one required a little, it seized hold of the pellet with its teeth or forceps, and detached a portion. The whole lump will not cleave off at once; but firmly adheres to the leg; from its tenacity, perhaps a string an inch long will be formed in separating, the piece obtained is immediately applied to their work, and the bee is ready to supply another with a portion; it doubtless gets rid of its load in this way; it is difficult to watch it till it is freed from the whole, as it is soon lost among its fellows. Now if this substance is not found in its natural state, how does it happen that they pack it on their legs just as they do when getting it from a board of an old hive, or pollen, when collected? They never take the trouble to pack the wax there, when elaborated. Do not these circumstances strongly favor the idea of its being a vegetable substance? Perhaps the reason of its being collected at this season in greater abundance, may be found in the fact, that the buds of trees and shrubs are now generally formed. Many kinds are protected from rain and frost, by a kind of gum or resinous coating. It may be found in many species of Populus, particularly the balsam poplar, (*Populus Balsamifera*) and the Balm of Gilead, (*Populus Candicans*). By boiling the buds of these trees, an aromatic resin or gum may be obtained, (used sometimes for making salve;) the odor is very similar to that emitted by propolis, when first gathered by the bees, or by heating it afterwards. In the absence of facts, we are apt to substitute theory. This appears to me to be very plausible. Yet I am ready to yield it as soon as facts decide differently. Perhaps not one bee in a thousand is engaged in collecting this substance—there being so few may be one reason why they are not often detected, yet few as they are, a few of us should set about close observation; something certain might decide. Apiarian science is sadly neglected; a large amount of error is mixed up with truth, that patient, scrutinizing investigation must separate.

REMARKS.

I feel anxious to get to the practical part of this work, which I hope will interest some readers who care but little about the natural history. I shall begin with spring, and will now endeavor to mix more of the practical with it, as we proceed to the end of the year. In order to illustrate some points of practice, I may have occasion to repeat some things already mentioned.

CHAPTER VII.

THE APIARY.

ITS LOCATION.

In the location of the apiary, one important consideration is, that it is convenient to watch in the swarming season; that the bees may be seen at any time from a door or window, when a swarm rises, without the trouble of taking many steps to accomplish it; because if much trouble is to be taken, it is too often neglected. Also, if possible, the hives should stand where the wind will have but little effect, especially from the northwest. If no hills or building offer a protection, a close, high board fence should be put up for the purpose. It is economy to do it—bees enough may be saved to pay the expense. During the first spring months, the stocks contain fewer bees than at any other season. It is then that a numerous family is important, for the purpose of creating animal heat to rear the brood, if for nothing else. One bee is of more consequence now than a half dozen in midsummer. When the hive stands in a bleak place, the bees returning with heavy loads, in a high wind, are frequently unable to strike the hive, and are blown to the ground; become chilled, and die. A chilly south wind is equally fatal, but not so frequent. When protected from winds, the hives may front any point you choose; east or south is generally preferred. A location near ponds, lakes, large rivers, &c., will be attended with some loss. Hard winds will fatigue the bees when on the wing, often causing them to alight in the water; where it is impossible to rise again until wafted ashore, and then, unless in very warm weather, they are so chilled as to be past the effort. I do not mention this to discourage any one from keeping them, when so situated, because some few must keep them thus or not at all. I am so situated myself. There is a pond of four acres, some twelve rods off. In spring, during high winds, a great many may be found drowned, and driven on shore. Although we cannot miss so few from a stock, it is nevertheless a loss as far as it goes.

DECIDE EARLY.

Whatever location is chosen, it should be decided upon as early in the spring as possible; because, when the chilling winds of winter have ceased for a day, and the sun, unobstructed, is sending his first warm rays to a frozen earth, the bees that have been inactive for months, feel the cheering influence, and come forth to enjoy the balmy air. As they come from their door, they pause a moment to rub their eyes, which have long been obscured in darkness.

BEES MARK THEIR LOCATION ON LEAVING THE HIVE.

They rise on the wing, but do not leave in a direct line, but immediately turn their heads towards the entrance of their tenement, describing a circle of only a few inches at first, but enlarge as they recede, until an area of several rods have been *viewed and marked*.

CHANGING STAND ATTENDED WITH LOSS.

After a few excursions, when surrounding objects have become familiar, this precaution is not taken, and they leave in a direct line for their destination, and return by their way-marks without difficulty. Man with his reason is guided on the same principles. There are a great many people who suppose the bee knows its hive by a kind of instinct, or is attracted towards it, like the steel to the magnet. At least, they act as if they did; as they often move their bees a few rods, or feet, after the location is thus marked, and what is the consequence? The stocks are materially injured by loss of bees, and sometimes entirely ruined. Let us trace the cause. As I remarked, the bees have marked the location. They leave the hive without any precaution, as surrounding objects are familiar. They return to their old stand and find no home. If there is more than one stock, and the removal has been from four to twenty feet, some of the bees may find a hive, but just as liable to enter the wrong one as the right. Probably they would not go over twenty feet, and very likely not that, unless the new situation was very conspicuous. If a person had but one stock, very likely the loss would be less, as every bee finding a hive, would be sure to be home, and none killed, as is generally the case when a few enter a strange hive.

CAN BE TAKEN SOME DISTANCE.

When bees are taken beyond their knowledge of country, some two miles or more, the case seems to be somewhat different, but not always without loss, especially if many hives are set too close. They leave the hive of course without knowing that the situation has been changed; perhaps get a few feet before strange objects warn them of the fact. When they return, the immediate vicinity is strange, and they often enter their neighbors' domicil.

DANGER OF SETTING STOCKS TOO CLOSE.

A case in point occurred in the spring of '49. I sold over twenty stocks to one person. He had constructed a bee-house, and his arrangement brought the hives within four inches of each other. The result was, he entirely lost several stocks; some of them were the best; others were materially injured, yet he had a few made better by the addition of bees from other hives; (sometimes a stock will allow strange bees to unite with them, but it is seldom, unless a large number enters—it is safest to keep each family by itself, under ordinary circumstances). These stocks, before they were moved, had been collecting pollen, and had their location well marked. Had they been placed six feet apart, instead of four inches, he probably would not have lost any, or even two feet might have saved them. I have often moved them at this season, and placed them at three feet distance, and had no bad results.

Facts like the foregoing, satisfied me long since that stocks should occupy their situation for the summer, as early as possible in the spring, at least before they mark the location; or if they must be moved after that, let it be nothing short of a mile and a half, and plenty of room between the hives.

SPACE BETWEEN HIVES.

As regards the distance between hives generally, I would say let it be as great as convenience will allow. Want of room makes it necessary sometimes to set them close; where such necessity exists, if the hives were dissimilar in color, some dark, others light, alternately, it would greatly assist the bees in knowing their own hive. But it should be borne in mind, that whenever economy of space dictates less than two feet, there are often bees enough lost by entering the wrong hive, which, if saved, would pay the rent of a small addition to a garden, or bee-yard. I have several other reasons to offer for giving plenty of room between hives, which will be mentioned hereafter.

SMALL MATTERS.

The reader who is accustomed to doing things on gigantic principles, will consider this long "yarn" about saving a few bees in spring, a rather small affair, and so it is; yet small matters must be attended to if we succeed; "a small leak will sink a ship." A grain of wheat is a small matter; 'tis only in the aggregate that its importance is manifest. The bee is small, the load of honey brought home by it is still less, and the quantity secreted in the nectary of each flower, yet *more minute*. The patient bee visits each, and obtains but a tiny morsel; by perseverance a load is obtained, and deposited in the hive; it is only by the accumulation of such loads that we find an object worthy our notice: here is a lesson; look to little things, and the manner in which they are multiplied, and preserved. It is much better to save our bees than waste them, and wait for others to be raised; "a penny saved is worth two-pence earned." If a stock is lost by small means, a corresponding effort is only necessary to save it. This trifling care is sometimes neglected through indolence. But I hope for better things generally; I am willing to believe it is thorough ignorance, not knowing what kind of care is necessary—how, when, and where to bestow it. This is what now appears to be my duty to tell. You will now sufficiently understand the cause of loss on this point; therefore, let it be a rule to have all ready in spring, before the bees leave their hives—the stands, bee-house, etc., and not change them.

ECONOMY.

If we keep bees for ornament, it would be well to build a bee-house, paint the hives, &c.; but as I expect the majority of readers will be interested in the profit of the thing, I will say that the bees will not pay a cent towards extra expenses; they will not do a whit more labor in a painted house, than if it was thatched with straw. When profit is the only object, economy would dictate that labor shall be bestowed only where there will be a remuneration.

CHEAP ARRANGEMENT OF STANDS.

So many kinds of bee-houses and stands have been recommended—all so different from what I prefer, that I perhaps ought to feel some hesitancy in offering one so cheap and simple; but as profit is my object, I shall offer no other apology. I have fifteen years' experience to prove its efficacy, and have no fears on this score in recommending it. I make stands in this way: a board about fifteen inches wide is cut off two feet long; a piece of chestnut or other wood, two inches square, is nailed on each end; this raises the board just two inches from the earth, and will project in front of the hive some ten inches, making it admirably convenient for the bees to alight before entering the hive, (when the grass and weeds are kept down, which is but little trouble). A separate piece for each hive is better than to have several on a bench together, as there can then be no communication by bees running to and fro. Also we are apt to give more room between them; and a board or plank will make a stand for as many stocks when cut in pieces, as if left whole; (and it ought to make more).

CANAL BOTTOM-BOARD DISCARDED.

I used what is termed a canal bottom-board, until I found out it did not pay expense, and have now discarded it, and succeed just as well. It is generally recommended as a preventive of robberies, and keeping out the moth. It may prevent one hive in fifty from being robbed; but as for keeping out the moth, it is about as good an assistant for it as can be contrived. It is a place of great convenience for the worms to spin their cocoons, and some ingenuity of the apiarian is requisite to get at them.

SOME ADVANTAGE IN BEING NEAR THE EARTH.

I am aware that I go counter to most apiarians, in recommending the stands so near the earth; less than two or three feet between the bees and the earth, it is said, will not answer any way. Mr. Miner is very positive on this point, in his Manual. I ventured to suggest to him, that there was more against it in theory, than in practice, and gave him my experience. In less than two years from that time I visited him, and found his bees close to the earth. Experience is worth a dozen theories; in fact, it is the only test to be depended upon. I shall not urge the adoption of any rule, that I have not proved by my own practice. The objection raised, is dampness from the earth, when too near; I am unable to perceive the least bad effect. Now let us compare advantages and disadvantages a little farther. One hive or a row of hives suspended, or standing on a bench, two or three feet from the earth, when approached by the bees on a chilly afternoon, (and we have many such in spring,) towards evening, even if there is not much wind, they are very apt to miss the hive and bottom, and fall to the ground, so benumbed with cold, as to be unable to rise again, and by the next morning are "no use" whatever. On the other hand, if they are near the earth, with a board as described, there is no *possibility* of their alighting under the hive, and if they should come short, and get on the ground, they can always creep, long after they are too cold to fly, and are able, and often do enter the hive without the necessity of using their wings.

Enough may be saved in one spring, from a few hives, in this way, to make a good swarm, which taken from several is not perceived; yet, as much profit from them might be realized, as if they were a swarm by themselves. A little contrivance is all that is needed to save them. To such as *must* and *will* have them up away from the earth, I would say, do suggest some plan to save this portion of your best and most willing servants; have an alighting board project in front of the hive at least one foot, or a board long enough to reach from the bottom of the hive to the ground, that they may get on that, and crawl up to the hive. Do you want the inducement? Examine minutely the earth about your hives, towards sunset, some day in April, when the day has been fair, with some wind, and chilly towards night, and you will be astonished at the numbers that perish. Most of them will be loaded with pollen, proving them martyrs to their own industry and your negligence. When I see a bench three feet high and no wider than the bottom of the hive, perhaps a little less, and no place for the bees to enter but at the bottom, and as many hives crowded on as it will hold, I no longer wonder that "bee-keeping is all in luck;" the wonder is how they keep them at all. Yet it proves that, with proper management, it is not so very precarious after all.

The necessary protection from the weather, for stocks, is a subject that I have taken some pains to ascertain; the result has been, that the cheapest covering is just as good as any; something to keep the rain and rays of the sun from the top, is all sufficient. Covers for each hive, like the bottom-board, should be separate, and some larger than the top.

UTILITY OF BEE-HOUSES DOUBTED.

I have used bee-houses, but they will not pay, and are also discarded. They are objectionable on account of preventing a free circulation of air; also, it is difficult to construct them, so that the sun may strike the hives both in the morning and afternoon; which in spring is very essential. If they front the south, the middle of the day is the only time when the sun can reach all the hives at once; this is just when they need it least; and in hot weather, sometimes injurious by melting the combs. But when the hives stand far enough apart, on my plan, it is very easily arranged to have the sun strike the hive in the morning and afternoon, and shaded from ten o'clock, till two or three, in hot weather.

Notwithstanding our prodigality in building a splendid bee-house, we think of economy when we come to put our hives in, and get them *too close*. "Can't afford to build a house, and give them so much room, no how."

CHAPTER VIII.

ROBBERIES.

Robbing is another source of occasional loss to the apiarian. It is frequent in spring, and at any time in warm weather when honey is scarce. It is very annoying, and sometimes gets neighbors in contention, when perhaps neither is to blame, farther than ignorance of the matter.

NOT PROPERLY UNDERSTOOD.

A person keeping many hives must expect to be accountable for all losses in his neighborhood, whether they are lost by mismanagement or want of management. Many people suppose, if one person has but one stock, and another has ten, that the ten will combine for plundering the one. There are no facts, showing any communication between different families of the same apiary, that I can discover. It is true, when one family finds another weak and defenceless, possessing treasure, they have no conscientious scruples about carrying off the last particle. The hurry and bustle attending it seldom escape the notice of the other families; and when one hive has been robbed in an apiary, perhaps two-thirds of the other families, sometimes all, have participated in the plunder. One family, if it be large, is just as likely, and more so, to find a weak one among the ten, and commence plundering, as the other way.

IMPROPER REMEDIES.

Notwithstanding it is common to hear remarks like this, "I had a *first-rate* hive of bees," (when the fact was he had not looked particularly at his bees for a month, to know whether it was so or not, and if he had, very likely would not know,) "and Mr. A.'s bees began to rob them. I tried every thing to stop it; I moved them around in several places to prevent their finding the hive. It did no good; the first I knew they were all gone—bees, honey, and all! The bees all joined the robbers." Now the fact is, that not one *good* stock of bees in fifty, will ever be robbed, if let alone; that is, if the entrance is properly protected. This moving the hive was enough to ruin any stock; bees were lost at every change, until nothing was left but honey to tempt the robbers; whereas, if left on its stand, it might have escaped.

A great many remedies have been given me gratis, which, had one-half been followed, would have ruined them. The fact is, with many people, the remedies are often the cause of the disease. The most fatal is, to move them a few rods; another, to close the hive entirely, (very liable to smother them); or, break out some comb and set the honey to running. There are some charms that affect them but little any way. Probably there are but few bee-keepers able to tell at once, *when bees are being robbed*. It requires the closest scrutinizing observation to decide.

DIFFICULTY IN DECIDING.

There is nothing about the apiary more difficult to determine, nothing more likely than to be deceived. It is generally supposed, when a number are outside fighting, that it is conclusive that they are also robbing, which is seldom the case. On the contrary, a show of resistance indicates a strong colony, and that they are disposed to defend their treasures. I no longer have any fears for a stock that has courage to repel an attack.

WEAK FAMILIES IN MOST DANGER.

It is weak families, that show no resistance, where we find the most danger. In seasons of scarcity, all *good* stocks maintain or keep sentinels about the entrance, whose duty it appears to be to examine every bee that attempts to enter. If it is a member of the community, it is allowed to pass; if not, it is examined on the spot. It would seem that a password was requisite for admittance, for no sooner does a stranger-bee endeavor to get in, than it is known. If without necessary credentials, there is evidence enough against it. Each bee is a qualified jurist, judge, and executioner. There is no delay; no waiting for witnesses for defence. The more a bee attempts to escape, the more likely it will be to receive a sting, unless it succeeds. How strange bees are known, would be nothing but theory, if I should attempt to explain. Let it suffice that they are known.

THEIR BATTLES.

I will here describe some of their battles. I have in the spring frequently seen the whole front side of the hive covered with the combatants, (but for such hives I have no fears; they are able to defend themselves.) Several will surround one stranger; one or two will bite its legs, another the wings; another will make a feint of stinging, while another is ready to take what honey it has, when worried sufficient to make it willing. It is sometimes allowed to go after yielding all its honey, but at others, is dispatched with a sting, which is almost instantly fatal. A bee is killed sooner by a sting, than by any other means, except crushing. Sometimes a leg will tremble, for a minute; the legs are drawn close to the body; the abdomen contracts to half its usual size, unless filled with honey. I have known a pint accidentally to enter a neighboring stock, and be killed in five minutes. The only places the sting will penetrate a bee are the joints of the abdomen, legs, the neck, &c. I have occasionally seen one bee drag about the dead body of its victim, being unable to withdraw its sting from a joint in the leg. During the fight, if it be to keep off those in search of plunder, a few bees may be seen buzzing around in search of a place unguarded to enter the hive. If such is found, it alights and enters in a moment. At other times, when about to enter, it meets a soldier on duty, and is on the wing again in an instant. But another time it may be more unfortunate, and be nabbed by a policeman, when it must either break away, or suffer the penalty of insect justice, which is generally of the utmost severity.

BAD POLICY TO RAISE THE HIVES.

A great many apiarians raise their hives an inch from the board early in spring. They seem to disregard the chance it gives robbers to enter on every side. It is like setting the door of your own house open, to tempt the thief, and then complain of depravity.

Let it be understood, then, that all good stocks, under ordinary circumstances, will take care of themselves. Nature has provided means of defence, with instinct to direct its use. Non-resistance may do for highly cultivated intellect in man, but not here.

INDICATIONS OF ROBBERS.

We will now notice the appearance about a weak hive that makes no resistance, and show the result to be a total loss of the stock, without timely interference. Each robber, when leaving the hive, instead of flying in a direct line to its home, will turn its head towards the hive to mark the spot, that it may know where to return for another load, in the same manner that they do when leaving their hive in the spring. The first time the young bees leave home, they mark their location, by the same process. A few of these begin to hatch from the cells very early; in all good stocks, often before the weather is warm enough for *any to leave the hive*. Consequently, it cannot be too early for them at any time in spring. These young bees, about the middle of each fair day, or a little later, take a turn of flying out very thickly for a short time. The inexperienced observer would be very likely to suppose such stock very prosperous, from the number of inhabitants in motion. This unusual bustle is the first indication of foul play, and should be regarded with suspicion; yet it is not conclusive.

A DUTY.

It is the duty of every bee-keeper, who expects to succeed, to know which his weak stocks are; an examination some cool morning, can be made by turning the hive bottom up, and letting the sun among the combs. The number of inhabitants in them is easily seen. When weak, close the entrance, till there is just room for one bee to pass at once. The first real pleasant days, at any time before honey is obtained plentifully, a little after noon, look out for them to commence robbing. Whenever a weak stock is taken with what appears to be a fit of unusual industry, it is quite certain they are either robbers or young bees; the difficulty is to decide which. Their motions are alike, but there is a little difference in color—the young bees are a shade lighter; the abdomen of the robbers, when filled with honey, is a little larger. It requires close, patient observation, to decide this point, and when you have watched close enough to detect this difference, you can decide without trouble.

A TEST.

But while you are learning this nice distinction, your bees may be ruined. We will, therefore, give some other means of protection.

Bees, when they have been stealing a sack of honey from a neighboring hive, will generally run several inches from the entrance before flying: kill some of these; if filled with honey, they are robbers; because it is very suspicious, to be filled with honey when leaving the hive; or sprinkle some flour on them as they come out, and have some one watch by the others to see if they enter. Another way is less trouble, but will take longer, before they are checked, if robbing. Visit them again in the course of half an hour or more, after the young bees have had time to get back, (if it should happen to be them); but if the bustle continues or increases, it is time to interfere. When the entrance has been contracted as directed, close it entirely till near sunset. When it has been left without, it should now be done, (giving room for only one bee at a time). This will allow all that belong to the hive to get in, and others to get out, and materially retard the progress of the robbers.

ROBBING USUALLY COMMENCES ON A WARM DAY.

Unless it should be cool, they will continue their operations till evening. Very often some are unable to get home in the dark, and are lost. This, by the way, is another good test of robbing. Visit the hives every warm evening. They *commence* depredations on the warmest days; seldom otherwise. If any are at work when honest laborers should be at home, they need attention.

REMEDIES.

As for remedies, I have tried several. The least trouble is to remove the weak hive in the morning to the cellar, or some dark, cool place, for a few days, until at least two or three warm days have passed, that they may abandon the search. The robbers will then probably attack the stock on the next stand. Contract the entrance of this in accordance with the number of bees that are to pass. If strong, no danger need be apprehended; they may fight, and even kill some; perhaps a little chastisement is necessary, to a sense of their duty.

COMMON OPINION.

There is an opinion prevalent that robbers often go to a neighboring stock, kill off the bees first, and then take possession of the treasures. To corroborate this matter, I have never yet discovered one fact, although I have watched very closely. Whenever bees have had all their stores taken, at a period when nothing was to be had in the flowers, it is evident they must starve, and last but a day or two before they are gone. This would naturally give rise to the supposition that they were either killed, or gone with the robbers.

A CASE IN POINT.

I have a case in point. Having been from home a couple of days, I found, on my return, a swarm of medium strength, that had been carelessly exposed, had been plundered of about fifteen pounds of honey, every particle they had.[13] About the usual number of bees were among the combs, to all appearance, very disconsolate. I at once removed them to the cellar, and fed them for a few days. The other bees gave over looking for more plunder, in the meantime. It was then returned to the stand, entrance nearly closed, as directed, &c. In a short time it made a valuable stock; but had I left it twenty-four hours longer, it probably would not have been worth a straw.

FURTHER DIRECTIONS.

When a stock has been removed, if the next stand contains a weak, instead of a strong one, it is best to take that in also; to be returned to the stand as soon as the robbers will allow it. If a second attack is made, put them in again, or if practicable, remove them a mile or two out of their knowledge of country; they would then lose no time from labor. Where but few stocks are kept, and not more than one or two stocks are engaged, sprinkle a little flour on them as they leave, to ascertain which the robbers are; then reverse the hives, putting the weak one in the place of the strong, and the strong one in the place of the weak one. The weak stock will generally become the strongest, and put a stop to their operations; but this method is often impracticable in a large apiary; because several stocks are usually engaged, very soon after one commences, and a dozen may be robbing one. Another method is, when you are *sure* a stock is being robbed, take a time when there are as many plunderers inside as you can get, and close the hive at once, (wire-cloth, or something to admit air, and at the same time confine the bees, is necessary;) carry in, as before directed, for two or three days, when they may be set out. The strange bees thus enclosed will join the weak family, and will be as eager to defend what is now *their* treasure, as they were before to carry it off. This principle of forgetting home and uniting with others, after a lapse of a few days, (writers say, twenty-four hours is sufficient for them to forget home) can be recommended in this case. It succeeds about four times in five, when a proper number is enclosed. Weak stocks are strengthened in this way very easily; and the bees being taken from a number of hives, are hardly missed. The difficulty is, to know when there are enough to be about equal, to what belongs to the weak stock; if too few are enclosed, they are surely destroyed.

COMMON CAUSE OF COMMENCING.

After all, bees being robbed is like being destroyed by worms; a kind of secondary matter; that is, not one strong stock in a hundred will ever be attacked and plundered on the first onset. Bees must be first tempted, and rendered furious by a weak hive; a dish of refuse honey set near them is sometimes sufficient to set them at work, also where they have been fed and not had a full supply. After they have once commenced, it takes an astonishing quantity to satiate their appetite. They seem to be perfectly intoxicated, and regardless of danger; they venture on to certain destruction! I have known a few instances where good stocks by this means were reduced, until they in turn fell a prey to others. I have for several years kept about one hundred stocks away from home, where I could not see them much, to prevent robbing. Yet I never lost a stock by this cause. I simply keep the entrance closed, except a passage for the bees at work during spring. It is true I have lost a few stocks, when the other bees took the honey, but they would have been lost any way.

SPRING THE WORST TIME.

As I before remarked in the commencement of this chapter, bees will plunder and fight at any time through the summer, when honey cannot be collected; but *spring* is the only time that such desperate and persevering efforts are made to obtain it. It is the only time the apiarian can be excused for having his hives plundered, or letting them stand in a situation for it. We then often have families reduced in winter and spring, from various causes, and when protected through this season, generally make good stocks. It is then we wish them to form steady, industrious habits, and not live by plunder. Prevention is better than cure; evil propensities should be checked in the beginning. The bee, like man, when this disposition has been indulged for a time, it is hard breaking the habit; a severe chastisement is the only cure; they too go on the principle of much wanting more.

NO NECESSITY TO HAVE THE BEES PLUNDERED IN THE FALL.

The apiarian having his bees plundered in the fall, is not fit to have charge of them; their efforts are seldom as strong as in spring, (unless there is a general scarcity,) the weak hives are usually better supplied with bees, and consequently a less number is exposed; but yet, when there are some very weak families, these should be taken away as soon as the flowers fail, or strengthened with bees from another hive. Particulars in fall management.

I have sometimes made my swarms equal, early in spring, by the following method, and I have also failed. Bees, when wintered together in a room, will seldom quarrel when first set out. When one stock has an over supply of bees, and another a very few, the next day or two after being out, I change the weak one to the stand of the strong one, (as mentioned a page or two back,) and all bees that have marked the location return to that place. The failure is, when too many leave the strong stock, making that the weak one, when nothing is gained. If it could be done when they had been out of the house just long enough for the proper number to have marked the location, success would be quite certain. But before an exchange of this kind is made, it would be well, if possible; to ascertain what is the cause of a stock being weak; if it is from the loss of a queen, (which is sometimes the case,) we only make the matter worse by the operation. To ascertain whether the queen be present, do not depend on the bees carrying in pollen; as most writers assert they will not, when the queen is gone; because I have *known* them do it so many times without, that I can assure the reader again, it is no test whatever. The test given in chapter III. page 73, is always certain.

CHAPTER IX.

FEEDING.

SHOULD BE A LAST RESORT.

Feeding bees in spring is sometimes absolutely necessary; but in ordinary seasons and circumstances, it is somewhat doubtful if it is the surest road to success, for the apiarian to attempt wintering any stock so poorly supplied with honey, that he feels satisfied will need feeding in the spring or before. I will recommend in another place (in fall management) what I consider a better disposition of such light families. But as some stocks are either robbed, or from some other cause, consume more honey than we expect, a little trouble and care may save a loss. Also bees are often fed at this season to promote early swarming, and filling boxes with surplus honey.

CARE NEEDED.

Considerable care is requisite, and but few know how to manage it properly. Honey fed to bees, is almost certain to get up quarrels among them. Sometimes strong stocks scent the honey given to weak ones, and carry it off as fast as supplied.

APPARENT CONTRADICTION WHEN FEEDING CAUSING STARVATION.

It is possible that feeding a stock of bees in spring, may cause them to starve! whereas, if let alone, they might escape. Notwithstanding this looks like a contradiction, I think it appears reasonable. Whenever the supply of honey is short, probably not more than one egg in twenty which the queen deposits, will be matured—their means not allowing the young brood to be fed. This appears from the fact that several eggs may be found in one cell. I transferred over twenty stocks in March, 1852—most of the cells occupied with eggs contained a plurality; two, three, and even four, were found in one cell; it is evident that all could not be perfected. Also, the fact of these eggs being at this season on the bottom-board. Now suppose you give such a stock two or three pounds of honey, and they are encouraged to feed a large brood, and your supply fails before they are half grown. What are they to do? destroy the brood and lose all they have fed, or draw on their old stores for a small quantity to help them in this emergency, and trust to chance for themselves? The latter alternative will probably be adopted, and then, without a timely intervention of favorable weather, the bees starve. The same effect is sometimes produced by the changes of the weather; a week or two may be very fine and bring out the flowers in abundance—a sudden change, perhaps frost, may destroy all for a few days. This makes it necessary to use considerable vigilance, as these turns of cold weather (when they occur) make it unsafe, till white clover appears; but if the spring is favorable, there is but little danger, unless they are robbed. If you take the necessary care about worms, you will know which are light, and which heavy, unless your hives are suspended; even then, it is a duty to know their true condition, in this respect. This is another advantage of the *simple* hive; merely raising one edge to destroy worms, tells you something about the honey on hand. To be very exact, the hive should be weighed when ready for the bees, and the weight marked on it; by weighing at any time after, tells at once within a few pounds of what honey there is on hand. Some allowance must be made for the age of the combs, the quantity of brood, &c. It is wrong to begin to feed without being prepared to continue to do so, as the supply must be kept up till honey is abundant.

HOW LONG IT WILL DO TO WAIT BEFORE FEEDING.

If it is wished to wait as long as possible, and not lose the bees, a test will be necessary to decide how long it will do to delay feeding. In this case, *strict attention will be necessary; they will need examination every morning.* If a light tap on the hive is answered b; a brisk, lively buzzing, they are not suffering yet; but if no answer is returned to your inquiry, it indicates a want of strength. Extreme destitution destroys all disposition to repel an attack. Sometimes a part of the bees will be too weak to remain among the combs, and will be lying on the bottom, and some few outside. If the weather is cool, they appear to be lifeless; yet they can be revived, and now *must he fed.*

DIRECTIONS FOR FEEDING.

Those among the combs may be able to move, though feebly. When this is the condition of things, invert the hive, gather up all the scattered bees, and put them in. Get some honey; if candied, heat it till it dissolves; comb honey is not so good without mashing; if no honey is to be had, brown sugar may be taken instead; add a little water, and boil it till about the consistence of honey, and skim it; when cool enough, pour a quantity among the combs, directly on the bees; cover the bottom of the hive with a cloth, securing it firmly, and bring to the fire to warm up. In two or three hours they will be revived, and may be returned to the stand, providing the honey given is all taken up; on no account let any honey run out around the bottom. The necessity of a daily visit to the hives is apparent from the fact, that if left over for one day, in the situation just described, it will be too late to revive them. At night, if you have a box cover, such as I have recommended, you may open the holes in the top of the hive; fill a small baking dish with honey or syrup, and set it on the top; put in some shavings to keep the bees from drowning, or a float may be used if you choose; it should be made of some very light wood, very thin, and full of holes or narrow channels, made with a saw. At the commencement of feeding, a few drops should be scattered on the top of the hive and trailed to the side of the dish, to teach them the way; after feeding a few times, they will know the road. When the weather is warm enough for them to take it during the night, it is best to feed at evening,—from four to eight ounces daily, is sufficient. If the family is very small, what honey is left in the morning may attract other bees; it is then best to take it out, or carry the hive in the house to a dark room, sufficiently warm, and feed them enough to last several days, and then return them to the stand; keeping a good lookout that they are not plundered, and again in a starving condition, until flowers produce honey sufficient.

WHOLE FAMILIES MAY DESERT THE HIVE.

When you have the means to keep up a supply of food, and time requisite to make feeding secure, perhaps it would not be advisable to wait till the last extremity before feeding, as a small family will sometimes entirely desert the hive, when destitute, if it occurs before they have much brood. In these cases, they issue precisely as a swarm; after flying a long time, they either return, or unite with some other stock. If they return, they need attention immediately. You may be certain there is something wrong, let the desertion take place when it may; in spring it may be destitution, or mouldy combs; at other times the presence of worms, diseased brood, &c. By whatever cause it is produced, ascertain it, and apply the remedy.

OBJECTIONS TO GENERAL FEEDING.

I have known it recommended, and practised by some apiarians, to feed bees all at once in the open air, in a large trough; but whoever realizes much profit by this method, will be very fortunate, as every stock in the neighborhood will soon scent it out, and carry off a good share, and nearly every stock at home will be in contention, and great numbers killed; the moment the honey is out, their attention is directed to other stocks. Another objection to this general feeding is, that some stocks are not necessitated at all, while others need it; but the strongest stock is pretty sure to get the most. NOW, as I cannot afford to divide with my neighbors in this way of feeding, and I suppose but few will be found who are willing to do it, I will give my method, which, when once arranged, is but little trouble.

ARRANGEMENT FOR FEEDING.

I got a tinman to make some dishes, two inches deep, 10×12 inches square, and perpendicular sides. A board was then got out, fifteen inches wide, and two feet long; two inches from one end, a hole is cut out the longest way, just the size of the dish, so that it will set in just even with the upper side of the board; a good fit should be made, so that no bees can get in around it; cleats should be nailed on the under side of the board, some over an inch thick, to prevent crowding the dish out. This is to go directly under the hive, but it is not ready yet, because if such dish is filled with honey under a hive, the bees would drown; if a float is put on to keep them out, it will settle to the bottom when the honey is out, and the bees cannot creep up the sides of tin very easily. Another thing, there is nothing to prevent the bees from making their combs to the bottom of this dish, two inches below the bottom of the hive; these things are to be prevented. Get out two pieces of half-inch board, ten inches long, one to be two inches wide, the other one and a half inches. With a coarse or thick saw, cut channels in the side of the strips, one-fourth inch deep, three-eighths or half an inch apart, crosswise the whole length. You will then want a number corresponding to the places sawed, of very thin shingles, or strips, say one-eighth of an inch thick, and one and three-fourths wide, and nine and a half long; these are to stand edgewise in the dish; the first two are to hold them in the channels at the ends. The narrow one needs a block one-half inch square, nailed on each end; on the edge, a strip of wire cloth is then nailed on, making the whole width just two inches. This is now put in the dish, wire cloth at the bottom, two inches from one end; two pins to act as braces will keep it there; the other wide one is placed against the other end, and pressed down even with the top of the dish. The thin pieces are now slipped into the channels even with the top; it is now ready to go under the hive to be fed. Let the two-inch space project out on the back side of the hive. A narrow board should be provided, some more than two inches wide, to cover it. Let the hive stand close on this board; the hole in the side is sufficient for the passage of bees at work, till very hot weather. Thus you see that the hive covers all but the space behind, which the board covers, and not a strange bee can get at the honey, without entering the hole at the side, and passing through among the bees belonging to the hive, which they will not often do; if the family is numerous, it makes it as safe as feeding on the top; with this advantage, there are no bees in the way to interfere while pouring in the food. When the bees are to be fed, raise the board at the back and pour in the honey; the wire-cloth in the bottom prevents all bees from entering this space, at the same time will let the honey pass through directly under the bees, which will take it up quicker than from any other place that I can put it; they will work all night even when the weather is quite cool. This board and feeder can be taken out when done feeding, and put away till wanted again; if left under through the summer, it affords the worms a place rather too convenient to spin their cocoons, where they are not easily destroyed.

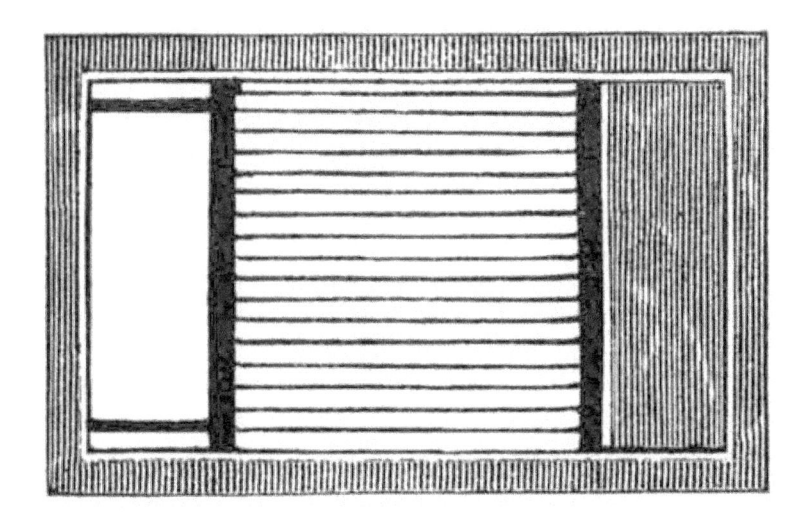

FEEDING TO INDUCE EARLY SWARMS.

If the object in feeding is to induce early swarms, of course the best stocks should be chosen for the purpose; but some care is necessary not to give too much, and fill the combs with honey, that ought to be filled with brood, and thereby defeat your object; one pound per day is enough, perhaps too much. The quantity obtained from flowers is a partial guide; when plenty, feed less; when scarce, more. Begin as soon as you can make them take it up in spring, and continue in accordance with the weather, till white clover blossoms, or swarms issue. Another object in feeding bees at this period, is to have the store combs all filled with inferior honey, so that when clover appears, (which yields our best honey,) there is no room except in the boxes to store it, which are now put on, and rapidly filled. When this last object is alone wished for, it is not much matter how much is given at a time, providing it is all taken up through the night; it will then take no time in day-light, when they might work on flowers; also, the bees would have no trouble in repelling any attempt of others to get at it.

WHAT MAY BE FED.

Inferior honey may be used for this purpose; Southern or West India is good, and costs but little. Even molasses sugar mixed with it will do; but they do not relish it so well when fed without the honey. I have usually taken about equal quantities of each, adding a pint of water to ten pounds of this mixture, and making it as hot as it will bear without boiling over, and skimming it.

IS CANDIED HONEY INJURIOUS?

There has an idea been advanced, that candied honey is injurious to bees, even said to be fatal. I never could discover any thing further, than it was a perfect waste, while in this state. When boiled, and a little water added, it appears to be just as good as any. Nearly every stock will have more or less of it on hand at this season; but as warm weather approaches, and the bees increase to warm the hive, it seems to get liquified, from this cause alone. The bees, when compelled to use honey from these cells, thus candied, waste a large portion; a part is liquid, and the rest is grained like sugar, which may be seen on the bottom-board, as the bees work it out very often. Another object in feeding bees, is to give inferior honey, mixed with sugar and flavored to suit the taste, to the bees, and let them store it in boxes for market. Now, I have no faith in honey undergoing any chemical change in the stomach of the bee,[14] and cannot recommend this as the honest course. Neither do I think it would be very profitable, feeding to this extent, under any circumstances. I have a few times had some boxes nearly finished and fit for market at the end of the honey season; a little more added would make them answer. I have then fed a few pounds of good honey, but always found that several pounds had to be given the bees to get one in the boxes.

CHAPTER X.

DESTRUCTION OF WORMS.

I shall not give a full history of the moth in this chapter, as spring is not the time they are most destructive. It will be further noticed under the head of Enemies of Bees. But as this is a duty belonging to spring, a partial history seems necessary.

As soon as the bees commence their labors, the worms are generally ready to begin theirs.

SOME IN THE BEST STOCKS.

You will probably find some in your best stocks; but don't be frightened; this is not the season when they often destroy your stocks, yet they injure them some.

HOW FOUND.

In the morning, when cool, raise the hive, and you will find them on the board. You must not suppose that these chaps are bred outside the hive, got their growth, and are now on their way among the bees, but the reverse. They are *bred in the hive*, and most of them are on the way out, and this is the precise time to arrest them and bring them to justice for their crimes.

A TOOL FOR THEIR DESTRUCTION.

I have used a simple tool, made in a few minutes, and very convenient in this business. Any one can make it. Get a piece of narrow hoop-iron, (steel would be better,) three-fourth inch wide, five inches long; taper from one side three inches from the end to a point; then grind each edge sharp; make three or four holes through the wide end, to admit small nails through it in the handle, which should be about two feet long and about half an inch square. Armed with this weapon, you can proceed. Raise the hive on one edge, and with the point of your sword you may pick a worm out of the closest corner, and easily scrape all from under the hive with it. Now, *be sure and dispatch every one*; not that the "little victim" will itself, personally, do much mischief; but through its descendants the mischief is to be apprehended. Very likely half of all you find will have finished their course of destruction, among the combs, and have voluntarily left them for a place to spin their cocoons. They are worried by the bees, if they are numerous, until satisfied that it is no safe place among them to make a shroud and remain helpless two or three weeks. Accordingly, when they get their growth they leave, get on the board on the bottom, become chilled and helpless in the morning, but again active by the middle of the day. Now, if they are merely thrown on the earth, a place there will be selected, if no better is found, for transformation; and a moth perfected ten feet from the hive is just as capable of depositing five hundred eggs in your hive, as if she had never left it.

Several generations are matured in the course of one summer: consequently, one destroyed at this season, may prevent the existence of thousands before the summer is over.

This is another subject of theoretical reasoning, and imposition, (at least in my opinion.) I wish the reader to judge for himself; get rid of whims and prejudice, and look at the subject candidly and fair; and if there is no corroborative testimony comes up to confirm any position that I assume, I shall not complain if my assertions fare no better than some others. Only defer judgment till you *know* for yourself.

Bees have ever received my especial regard and attention; and my enthusiasm may blind my judgment. I may be prejudiced, but will not be wilfully wrong. I have found so many theories utterly false, when carried out in practice, that I can depend on no one's hypothesis, however plausible, without facts in practice to support it. No one should be fully credited without a test. To return to our subject.

MISTAKEN CONCLUSIONS.

It is supposed by many, when these worms are found on the board, they get there by accident, having dropped from the combs above. They seem not to understand that the worm generally travels on safe principles; that is, he attaches a thread to whatever he travels over. To be satisfied on this point, I have many times carefully detached his foot-hold, when on the side of the hive or other place, where he would fall a few inches, and always found him with a thread fast at the place he left, to enable him to regain his position if he chose. Is it not probable, then, that whenever he leaves the combs for the bottom-board, he can readily ascend again? No doubt he often does, to be driven down again by the bees. Now, what I wish to get at by all this preamble, is simply this: that all our trouble and worrying to prevent the worms from again ascending to the combs—by wire hooks, wire pins, screws, nails, turned pins, clam-shells, blocks of wood, &c., is perfect nonsense, when half or more of them would not harm the bees any more if they did, and might as well go there as any where else. Besides, these useless "fixins" are very often a positive injury to the bees.

OBJECTIONS TO SUSPENDED BOTTOM-BOARD.

Suppose, if you please, that the worm has no thread attached above, and your board is far enough from the bottom of the hive to prevent his reaching it. Of course, he can't get up; but how are your bees to do any better? The worm can reach as high as they can. The bee can fly up, you think; so it will, sometimes; but will try a dozen times first to get up without, and when it does, it is a very bad position to start from, being a smooth board. In hot weather it does better. Did you ever watch by a hive thus raised, in April or May, towards night, when it was a little cool, and see the industrious little insects arrive with a load as heavy as they could possibly carry, all chilly, and nearly out of breath, scarcely able to reach home, and there witness their vain attempts to get among their fellows above them? If you never witnessed this, I wish you would take some pains for it, and when you find them giving up in despair, when too chilly to fly, and perishing after many fruitless attempts for life, I think, if you possess sympathy, benevolence, or even selfishness, you will be induced to do as I did—discard at once wire hooks and all else from under the hive in the spring, and give the bees, when they do get home with a load, under such circumstances, what they richly deserve, and that is, *protection*.

ADVANTAGE OF THE HIVE CLOSE TO THE BOARD.

An inch hole in the side of the hive, a few inches from the bottom, as a passage for the bees, is needed, as I shall recommend letting the hive close to the board; it is essential on account of robbing; also, it is necessary to confine as much as possible the animal heat, in most hives, during the season the bees are engaged in rearing young brood; and warmth is necessary to hatch the eggs, and develop the larvæ; we all know that when the hive is close, less heat will pass off than if raised an inch.

OBJECTION ANSWERED.

You object to this, and tell me, "the worms will get between the bottom of the hive and the board." Well, I think they will, and what then? Why I expect if you intend to succeed, that you will get them out, and crush their heads; if you cannot give as much attention as this, better not keep them, or let some one have the care of them that will. I am as willing to find a worm under the edge of the hive, and dispatch it, as to have it creep into some place out of sight, and change to the moth. I once trimmed off the bottom of my hives to a thin edge, so they did not have this place for their cocoons, but now prefer to have them square. *All profit* is seldom obtained with anything. If you plant a field with corn, you do not expect that the whole work for the crop is finished. Neither should you expect when you set up a stock of bees, that a full yield will be realized without something more. If you are remunerated by keeping the weeds from your corn, be assured it is equally profitable to weed out your bees.

INSUFFICIENCY OF INCLINED BOTTOM-BOARD.

Now do not be deceived in this matter, and through indolence be induced to get those hives with descending bottom-boards, to throw out the worms as they fall, and hope by that means to get rid of the trouble; (I have already, in another chapter, expressed doubts of this). But we will *now* suppose such descending bottom-boards capable of throwing every worm that touches it "heels over head" to the ground; what have we gained? His neck is not broken, nor any other *bone* of his body! As if nothing extraordinary had happened, he quietly gathers himself up, and looks about for snug quarters; he cares not a fig for the hive now; he gormandized on the combs until satisfied, before he left them, and is glad to get away from the bees any how. A place large enough for a cocoon is easily found, and when he again becomes desirous of visiting the hives, it is not to satisfy his own wants, but to accommodate his progeny; he is then furnished with wings ample to carry him to any height that you choose to put your bees.

A MOTH CAN GO WHERE BEES CAN.

A hive that is proof against the moth, is yet to be constructed. We frequently hear of them, but when they come to be tested, somehow these worms get where the bees are. When your hives become so full of bees, that they cover the board in a cool morning, the worms will be seldom found there, except under the edge of the hive.

TRAP TO CATCH WORMS.

You may now raise it, but you may still catch the worms by laying under the bees a narrow shingle, a stick of elder split in two lengthwise, and the pith scraped out, or anything else that will afford them protection from the bees, and where they may spin their cocoons. These should be removed every few days, and the worms destroyed, and the trap put back. Do not neglect it till they change to the moth, and you have nothing but to remove the empty cocoon.

BOX FOR WREN.

If you would take the trouble to put up a cage or two for the wren to nest in, he would be a valuable assistant in this department of your labor. He would be on the lookout when you were away, and many worms, while looking up a hiding-place in some corner, would be relieved from all further trouble by being deposited in his crop. The cage for him need not be more than four inches square; it may be fastened near as possible to the bees; to a post, tree, or side of some building a few feet high. I have seen the skull of some animal (horse or ox) used, and is very convenient for them, the cavity for the brains being used for the nest. A person once told me the wren would not build in one that he had put up. On examination, the stake to support it was found driven into the only entrance. I mention this to show how little some people understand what they do. It is sometimes well enough to know why a thing is to be done, as to know it *must* be done. I could tell you to do a great many things, but then you would like to know *why*, then *how* to do it. Now if this prolixity is unnecessary for you, another may need it. You must remember I am endeavoring to teach some few to keep bees, who are not over supplied with ingenuity.

CHAPTER XI.

PUTTING ON AND TAKING OFF BOXES.

Putting on boxes may be considered a duty intermediate between spring and summer management. I cannot recommend putting them on as early as the last of April, or first of May, in ordinary circumstances. It is possible to find a case that it would be best. But before the hive is full of bees it is generally useless, very likely a disadvantage, by allowing a portion of animal heat to escape that is needed in the hive to mature the brood. Also, moisture may accumulate until the inside moulds, &c. Some experience and judgment is necessary to know about what time boxes are needed. That boxes *are needed* at the proper season, I think I shall not need an argument to convince any one, in the present day. Bee-keepers have generally discarded the barbarous practice of killing the bees to obtain the honey. Many of them have learned that a good swarm will store sufficient honey for winter, besides several dollars worth as profit in boxes.

ADVANTAGE OF THE PATENT VENDER.

Here is where the patent vender has taken the advantage of our ignorance, by pretending that no other hive but *his ever obtained such quantities, or so pure in quality*.

TIME OF PUTTING ON—RULE.

It is probable a great many readers will need the necessary observation to tell precisely when the hive is full of honey; it may be full of bees, and not of honey. And yet the only rule that I can give to be generally applied, is, when the bees begin to be crowded out, but a day or two before would be just the right time, that is, when they are obtaining honey—(for it should be remembered that they do not always get honey when beginning to cluster out). This guide will do in place of a better one, which close observation and experience only can give. By observing a glass hive attentively, in those cells that touch the glass on the edge of the combs, whenever honey is being deposited here abundantly, it is quite evident that the flowers are yielding it just then, and other stocks are obtaining it also. Now is the time, if any cluster out, to put on the boxes. When boxes are made as I have recommended, that is, the size containing 360 solid inches, it is advisable to put on only one at first; when this is full either of bees or honey, and yet bees are crowded outside, the other can be added. This is before swarming; too much room might retard the swarming a few days, but if crowded outside, it indicates want of room, and the boxes can make but little difference. It is better to have one box well filled than two half full, which might be the case if the bees were not numerous. The object of putting on boxes before swarming, is to employ a portion of the bees, that otherwise would remain idly clustering outside two or three weeks, as they often do, while preparing the young queens for swarming. But when all the bees can be profitably engaged in the body of the hive, more room is unnecessary.

MAKING HOLES AFTER THE HIVE IS FULL.

Whenever it is required to put boxes on a hive that has no holes through the top, it need not prevent your getting a few pounds of the purest honey that may be had, just as well as to have a portion of the bees idle. I always endeavor to ascertain in what direction the sheets of comb are made, and then mark off the row of holes on the top, at right angles with them.

ADVANTAGE OF PROPER ARRANGEMENT.

Two inches being nearly the right distance, each one will be so made that a bee arriving at the top of the hive between any two sheets will be able to find a passage into the box, without the task of a long search for it; which I can imagine to be the case when only one hole for a passage is made, or when the row of holes is parallel with the combs. A hive might contain eight or ten sheets of comb, and a bee desirous of entering the box might go up between any two, many times, before it found the passage. It has been urged that every bee soon learns all passages and places about the hive, and consequently will know the direct road to the box. This may be true, but when we recollect that all within the hive is perfect darkness—that this path must be found by the sense of feeling alone—that this sense must be its guide in all its future travels—that perhaps a thousand or two young workers are added every week, and these have to learn by the same means—it would seem, if we studied our own interest, we would give them all the facility possible for entering the boxes. What way so easy for them as to have a passage, when they get to the top, between each comb? That bees do not know all roads about the hive, can be partially proved by opening the door of a glass hive. Most of the bees about leaving, instead of going to the bottom for their exit, where they have departed many times, seem to know nothing of the way, but vainly try to get out through the glass, whenever light is admitted.

I am so well convinced of this, that I take some pains to accommodate them with a passage between each comb; they will then at least lose no time by mistakes between the wrong combs, crowding and elbowing their way back through a dense mass of bees which impede every step, until again at the top perhaps between the same combs, perhaps right, perhaps farther off than at first; when I suppose they try it again; as boxes are filled sometimes under just such circumstances.

To assist them as much as possible, when new hives are used for swarms, I wait till the hive is nearly filled before making the holes to ascertain the direction of the combs. We all know it is uncertain which way the combs will be built, when the swarm is put in, unless guide-combs are used.[15] When holes are made before the bees are put in, guide-combs as directed for boxes should be put in; (of course they should cross at right angles the row of holes).

DIRECTIONS FOR BORING HOLES IN FULL STOCKS.

To make holes in the top after the combs are made,—Mark out the top as directed for making hives and boxes. A centre bit or an auger bit with a lip or barb is best, as that cuts down a little faster than the chip is taken out, leaving it smooth; when nearly through, a pointed knife can cut the remainder of the chip loose, and it can be taken out; if it is between the combs, it is well; if directly over the centre of one, it is a little better; with the knife take out a piece as large as a walnut; even if honey is in it, no harm will be done. The bees will then have a passage through from either side of the comb.

After you have opened one hole, very likely the bees will want to see what is going on over head, and walk out to reconnoitre. To prevent their interference, use some tobacco-smoke, and send them down out of your way, till your hole is finished. Now lay over this a small stone or block of wood, and make the others in the same way. When all are done, blow in some smoke as you uncover them, and put on your box. This process is not half so formidable as it appears; I have in this way bored hundreds. You will remember my hives are not as high as many others keep them, they are in about as convenient a position as I can get them. This method saves me the trouble of sticking the guide-combs in my hives; also, the necessity of covering or stopping the holes. Dr. Bevan and some others have made a cross-bar hive, instead of nailing on a top in the usual way; a half-inch board of the right length is cut into strips, some over an inch wide, and half an inch apart, across the top. It is plain that in such a hive a bee can pass into the box whenever it arrives at the top, without difficulty. I will here repeat the objection to allowing too much room, to pass into the boxes, that you may see the disadvantages of the extremes of too little and too much room. In these cross-bar hives, the animal heat rises into the box from the main hive, making it as warm as below; the queen goes up with the bees, and finding it warm and convenient for breeding, deposits her eggs; and young brood as well as honey is found there. When we think it full, it is then indispensable to return it, if taken off, till they hatch, (otherwise they spoil it by moulding), which makes the combs dark, tough, &c. Another objection to such open tops is, that open bottom boxes must be used, which are not half as neat for market.

TO BE TAKEN OFF WHEN FILLED.

This advantage attends glass boxes: while being filled, the progress can be watched till finished, when they should be taken off to preserve the purity of the combs. Every day the bees are allowed to run over them, renders them darker. Consequently, when our bees are a long time filling a box, it is not as purely white as when filled expeditiously.

TIME TAKEN TO FILL A BOX.

Two weeks is the shortest time I ever had any filled and finished. This, of course, depends on the yield of honey, and size of the swarm; three or four weeks are usually taken for the purpose. I have before said that the first yield of honey nearly fails in this section, usually about the 20th of July; there are some variations, later or earlier, according to the season. In other places it may be much later.

WHEN TO TAKE OFF BOXES PART FULL.

It can be ascertained by occasionally raising the cover to your glass boxes. When no more is being added, all boxes that are worth the trouble should be taken off; if left longer the comb gets darker, and such cells of honey as are not sealed over, (and sometimes the majority are such,) the bees generally remove down into the hive.

TOBACCO SMOKE PREFERRED TO SLIDES.

When boxes are to be taken off, if a slide of tin, zinc, &c., is used to close the holes, some of the bees are apt to be crushed, others will find themselves minus a head, leg, or abdomen, and all of them be irritable for several days. A little tobacco smoke is preferable, as it keeps all quiet. Just raise the box to be taken off sufficient to puff under it some smoke, and the bees will leave the vicinity of the holes in an instant; the box can then be removed, and another put on if necessary, without exciting their anger in the least.

MANNER OF DISPOSING OF THE BEES IN THE BOXES.

Arouse the bees by striking the box lightly four or five times. If all the cells are finished, and honey is still obtained, turn the box bottom up, near the hive from which it was taken, so that the bees can enter it without flying; by this means you can save several young bees, that have never left the hive and marked the location, and a few others too weak to fly, but will follow the others into the hive; (such are lost when we are obliged to carry them at a distance.) Boxes can be taken off either in the morning or evening; if in the morning, it may stand several hours when the sun is not too hot, but on no account let it stand in the sun in the middle of the day, as the combs will melt. The bees will all leave, sometimes in an hour; at others they will not be out in three. They may be taken off at evening and stand till morning, in fair weather; if not too cool, they are generally all out; but here is some risk of the moth finding it and depositing her eggs; perhaps one in fifty may be thus found.

BEES DISPOSED TO CARRY AWAY HONEY.

When boxes are taken off at the end of the honey season, a different method of getting rid of the bees must be adopted, or we lose our honey. Unless the combs are all finished, we lose some then any way, as most of the bees fill themselves before leaving; they carry it home and return for more immediately, and take it all, if not prevented. It has been recommended to take it to some dark room with a small opening to let the bees out; in the course of the day they will sometimes all leave; but this method I have found unsafe, as they sometimes find the way back. When a large number of boxes are to be managed, a more expeditious mode is, to have a large box with close joints, or an empty hogshead, or a few barrels with one head out, set in some convenient place; put the boxes in, one above another, but not in a manner to stop the holes; over the top throw a sheet of one thickness, a thin one is best, as it will let through more light. The bees will leave the boxes, creep to the top, and get on the sheet; take this off and turn it over a few times; in this way all may be got rid of without the possibility of carrying off much honey. All that know the way will return to the hive, but a few young ones are lost.

NOT DISPOSED TO STING.

They seldom offer to sting during this part of the operation, even when the box is taken off without tobacco smoke, and carried away from the hive; after a little time, the bees finding themselves away from home, lose all animosity.

As honey becomes scarce, less brood is reared; a great many cells that they occupied are soon empty; also, several cells that contained honey have been drained, and used to mature the portion of brood just started at the time of the failure. We can now understand, or think we do, why our best stocks that are very heavy, that but a few days before were crowded for room and storing in boxes, are now eager for honey to store in the hive; as there is abundant room for several pounds. They will quickly remove to the hive the contents of any box left exposed; or even risk their lives by entering a neighboring hive for it; after being allowed to make a beginning, under such circumstances.

RULE.

During a yield of honey, take off boxes as fast as they are filled, and put on empty ones. At the end of the season take all off. Not one stock in a hundred will starve that has worked in boxes, that is, when the hive is the proper size, and full before adding the boxes, unless robbed or other casualty.

CHAPTER XII.

SECURING HONEY FROM THE MOTH.
TWO THINGS TO BE PREVENTED.

When the boxes are free from the bees, two things are to be prevented, if we wish to save our honey till cold weather. One is to keep out the worms, the other to prevent souring. The last may be new to many, but some few of us have had it caused by dampness in warm weather. The combs become covered with moisture, a portion of the honey becomes thin like water, and instead of the saccharine qualities we have the acid. Remedy: keep perfectly dry and cool, if you can, but dry at any rate.

APT TO BE DECEIVED ABOUT THE WORMS.

But the worms, you can surely keep them out, you think, since you can seal up the boxes perfectly close, preventing the moth or even the smallest ant from entering! Yes, you may do this effectually, but the worms will often be there somehow, unless in a very low temperature, such as a very cool cellar, or in house, and then you have dampness to guard against. I have a little experience in this matter that spoils your theory entirely. I have taken off glass jars, and watched them till the bees were all out, and was *certain the moth did not come near* them, then immediately sealed them up; absolutely preventing access afterwards, (I could do this with a jar more effectually than a box which is made of several pieces,) I then felt quite sure that I was ahead, and should have no trouble with the worms, as had often been the case before. I was sadly mistaken.

THEIR PROGRESS DESCRIBED.

In a few days, I could see at first a little white dust, like flour, on the side of the combs, and on the bottom of the jar. As the worms grew larger, this dust was coarser. By looking closely at the combs, a small white thread-like line was first perceptible, enlarging as the worm progressed.

When combs are filled with honey, they go only on the surface, eating nothing but the sealing of the cells; seldom penetrating to the centre, without an empty cell to give the chance. Disgusting as they seem to be, they dislike being daubed with honey. *Wax, and not honey, is their food.*

The reader would like to know how these worms came in the jars, when, to all appearance, it *was a physical impossibility.* I would like to tell positively, but cannot. But I will guess, if you will allow it. I will first premise, that I do not suppose they are generated spontaneously! Their being found there, then, would indicate some agent or means not readily perceived.

A SOLUTION OFFERED.

The hypothesis that I offer is original and new, and therefore open for criticism; if there is a better way to account for the mystery, I would be glad to know it.

From the first of June till late in the fall, the moth may be found around our hives, active at night, but still in the day. The only object probably is to find a suitable place to deposit its eggs, that the young may have food; if no proper and convenient place is found, why, I suppose it will take up with such as it *can* find; their eggs *must* be deposited somewhere, it may be in the cracks in the hive, in the dust at the bottom, or outside, as near the entrance as they dare approach. The bees running over them may get one or more of these eggs attached to their feet or bodies, and carry it among the combs, where it may be left to hatch. It is not at all probable that the moth ever passed through the hive among the bees, to deposit her eggs in the jars before mentioned. Had these jars been left on the hive, not a worm would have ever defaced a comb; because, when the bees are numerous, each worm as soon as it commences its work of destruction will be removed, that is, when it works on the surface, as in the boxes of honey—in breeding combs, they get in the centre and are more difficult to remove. By taking off these jars and removing the bees, it gave all the eggs that happened to be there a fair chance. Many writers finding the combs undisturbed when left on the hive till cold weather, recommend that as the only safe way, preferring to have the combs a little darker, than the risk of being destroyed by the worms. But I object to dark combs, and leaving the boxes will effectually prevent empty ones taking their places, which are necessary to get all the profits. I will offer a few more remarks in favor of my theory, and then give my remedy for the worms. I have found in all hives where the bees have been removed in warm weather, say between the middle of June and September, (and it has been a great many,) moth eggs enough among the combs to destroy them in a very short time, unless kept in a very cool place; this result has been uniform. Any person doubting this, may remove the bees from a hive that is full of combs in July or August; and close it to prevent the *possibility* of a moth entering, set it away in a temperature ranging from sixty to ninety, and if there are not worms enough to satisfy him that this is correct, he will have better success than I ever did. Yet, no such result will follow, when the bees are left among the combs, unless the swarm be very small; then the injury done will be in proportion. A strong stock may have as many moth eggs among the combs as a weak one, yet one will be scarcely injured, while the other may be nearly or quite destroyed.

Now, if this theory be correct, and the bees do actually carry these eggs among the combs, is there not a great deal of lost labor in trying to construct a moth-proof hive? The moth, or rather the worms, are ever present to devour the combs, whenever the bees have left them in this season.

METHOD OF KILLING WORMS IN BOXES.

Now, whether you are satisfied or not with the foregoing, we will proceed with the remedy. Perhaps you may find one box in ten that will have no worms about it, others may contain from one to twenty when they have been off a week or more. All the eggs should have a chance to hatch, which in cool weather may be three weeks. They should be watched, that no worms get large enough to injure the combs much, before they are destroyed. Get a close barrel or box that will exclude the air as much as possible; in this put the boxes, with the holes or bottom open. In one corner leave a place for a cup or dish of some kind, to hold some sulphur matches while burning. (They are made by dipping paper or rags in melted sulphur.) When all is ready, ignite the matches, and cover close for several hours. A little care is required to have it just right: if too little is used, the worms are not killed; if too much, it gives the combs a green color. A little experience will soon enable you to judge. If the worms are not killed on the first trial, another dose must be administered. Much less sulphur will adhere to paper or rags, if it is very hot, when dipped, than when just above the temperature necessary to melt it; this should be considered, as well as the number of boxes to be smoked, size of the vessel used in smoking them, &c.

Whether this gas from burning sulphur will destroy the eggs of the moth before the worm appears, I have not tested sufficiently to decide; but I do know that it is an effectual quietus for the larvæ!

FREEZING DESTROYS THEM.

Boxes taken off at the end of warm weather, and exposed in a freezing situation through the winter, appear to have all the worms as well as eggs for them destroyed by the cold; consequently, all boxes so exposed, may be kept any length of time; the only care being necessary, to shut out the moth effectually. But don't forget to look out for all combs from which the bees have been removed in warm weather. I prefer taking off all boxes at the end of the first yield of honey, even when I expect to put them on again for buckwheat honey. The bees at this season collect a great abundance of propolis, which they spread over the inside of the boxes as well as hive; in some instances it is spread on the glass so thick as to prevent the quality of honey being seen. There is no necessity for boxes on a hive at any season when there is no yield of honey to fill them. Sometimes even in a yield of buckwheat honey, a stock may contain too few bees to fill boxes, but just a few may go into them and put on the propolis; this should not be allowed, as it makes it look bad when used another year. At this season, (August) some old stocks may be full of combs, and but few bees, but swarms when they have got the hive full in time, are very sure to have bees enough to go into the boxes to work. I have known them to do so in three weeks after being hived.

OBJECTION TO USING BOXES BEFORE THE HIVE IS FULL.

Some put on boxes at the time of hiving the bees. In such cases the box is often filled first, and nearly as often will contain brood. I consider it no advantage, and often a damage to do so; as I want the hive full any way—and then if they have time let them into boxes, although it may be buckwheat, instead of clover honey that we get.

CHAPTER XIII.

SWARMING.

TIME TO EXPECT THEM.

The season for regular swarms in this section, I have known to commence the 15th of May, and in some seasons the 1st of July. The end is about the 15th of the latter month, with some exceptions. I have had one as late as the 21st; also a few buckwheat swarms between the 12th and 25th of August.

The subject now before us is one of thrilling interest. To the apiarian the prospect of an increase of stocks is sufficient to create some interest, even when the phenomenon of swarming would fail to awaken it. But to the naturalist this season has charms that the indifferent beholder can never realize.

ALL BEE-KEEPERS SHOULD UNDERSTAND IT AS IT IS.

As a guide in many cases, it is important that the practical apiarian should understand this matter *as it is*, and not as said to be by many authors. I shall be under the necessity of differing from nearly all in many points.

MEANS OF UNDERSTANDING IT.

This is another case of "when doctors disagree, who shall decide?" You, reader, are just the person. There is no need of a doctor at all in this matter. I will endeavor to give a test for most of my assertions. To make this subject as plain as possible in this place, I may repeat some things said before. The facts related have come under my own observation. I have probably taken more pains than most bee-keepers, to understand this matter to the bottom *from the beginning*, (I mean the bottom of the cells). But few apiarians have made the number of examinations that I have to get at the *modus operandi* of swarming. Perhaps I ought not to expect full credit for veracity, when I assure the reader that I have inverted more than one hundred stocks to get a peep at the royal cells, some of them near a dozen times in one summer. I have inverted them frequently for the purpose of obtaining cells. But generally to see when such cells are being made, when they contain eggs, when these eggs are sufficiently matured for swarming, or abandoned and destroyed, &c.

By these signs I predict with certainty (almost) when to expect swarms, and when to cease looking for them.

INVERTING A STOCK RATHER FORMIDABLE AT FIRST.

To a person that has never inverted a hive full of bees, even to overflowing, or never has seen it done, it appears like a great undertaking, as well as the probability of ruining the stock! But after the first trial, the magnitude of the performance is greatly diminished, and will grow less with every repetition of the feat, until there is not the least dread attending it. Without tobacco smoke I hardly think it practicable, but with it, there is not the least difficulty. It would be very unsatisfactory to turn over a hive and nothing to drive the bees away from the very places on the combs that you wish particularly to inspect. The smoke is just the thing to do it! As for the bad effects of such overturning and smoking, I never discovered any.

REQUISITES BEFORE PREPARATION OF QUEEN'S CELLS.

I have found the process for all regular swarms something like this: before they commence, two or three things are requisite. The combs must be crowded with bees; they must contain a numerous brood advancing from the egg to maturity; the bees must be obtaining honey either by being fed or from flowers. Being crowded with bees in a scarce time of honey is insufficient to bring out the swarm, neither is an abundance sufficient, without the bees and the brood. The period that all these requisites happen together, and remain long enough, will vary with different stocks, and many times do not happen at all through the season, with some.

These causes then appear to produce a few queen-cells, generally begun before the hive is filled, (sometimes when only half full, but usually remain as rudiments till the next year, when the foregoing conditions of the stock may require their use).

STATE OF QUEEN'S CELL WHEN USED.

They are about half finished, when they receive the eggs; as these eggs hatch into larvæ, others are begun, and receive eggs at different periods for several days later. The number of such cells seem to be governed by the prosperity of the bees: when the family is numerous, and the yield of honey abundant, they may amount to twenty, at other times perhaps not more than two or three; although several such cells may remain empty. I have already said that a failure, (or even a partial one), in the yield of honey at any time from the depositing of royal eggs till the sealing of the cells, (which is about ten days), would be likely to bring about their destruction. Even after being sealed, I have found a few instances where they were destroyed.

STATE WHEN SWARMS ISSUE.

But when there is nothing precarious about the honey, the sealing of these cells is the time to expect the first swarm, which will generally issue the first fair day after one or more are finished. I never missed a prediction for a swarm 48 hours, when I have judged from these signs, in a prosperous season. When there is a partial failure of honey, the swarm sometimes will wait several days after finishing them.

CLUSTERING OUTSIDE NOT ALWAYS TO BE DEPENDED UPON.

The clustering out of the bees I find but a poor criterion to judge from, further than full hives do swarm—many such do not.

EXAMINATIONS—THE RESULT.

I will detail a few circumstances, that have led to these conclusions. Some years ago the honey began to fail, when only about one third of my good stocks had cast swarms; and all at once, the issues began to "be few and far between." I had previously examined, and found they had gone into preparations pretty extensively; by having not only constructed cells, but occupied them with royal eggs and larvæ. Now I examined again, and found five out of six had destroyed them, (at the same time the bees clustered out extensively). This put an end to all hopes of swarms here. Some few had finished their cells, and these, I had some hopes, would send out the swarms; but the dry weather caused some misgivings. After waiting three or four days and none coming, I found these sealed cells destroyed also, and had no more swarms that season. Subsequent observations have fully confirmed these things. One season some of the hives commenced preparations at two different periods, and then abandoned them without swarming at all, through the summer. The first time it was the last of May, the next in July.

REMARKS.

The failure of honey was the cause, without any doubt. And who shall say, these bees were not wise in their conduct? What prudent man would emigrate with a family, if the prospect of a famine was plainly indicated, when, by remaining at home, there was enough, at least for the present? Who can help but admire this wise and beautiful arrangement? The combs must contain brood; the bees must find honey during the rearing of the queens. If a swarm were to issue the moment of obtaining honey, the consequence might be fatal, as there would not be a numerous brood to hatch out, and replenish the old stock with bees sufficient to keep out the worms. Were they to issue at any time, as soon as the bees had increased enough in numbers to spare a swarm, without regard to the yield of honey, they might starve.

CONFLICTING THEORIES.

I find many theories conflicting with these views, which appear to call for some remarks. It is generally supposed that a young queen must be matured to issue with the swarms, and the old one with the old bees are permanent residents of the old hive.

BOTH OLD AND YOUNG LEAVE WITH SWARMS.

It is probable that no rule governs the issue of workers. Old and young come out promiscuously. That old bees come out may be known sometimes, by so many leaving, that not a quarter as many will be left, as commenced work in the spring. That young bees leave, any one may be satisfied on seeing a swarm issue; a great many too young and weak to fly will drop down in front of the hive, having come out now for the first time, and perhaps some of them had not been out of the cell an hour; these very young bees are known by the color.

CAUSE OF THE QUEEN'S INABILITY TO FLY SUGGESTED.

The old queen often gets down in the same way; but I would assign another cause for her inability to fly; that is, I would suggest it to be her burden of eggs.

EVIDENCE OF THE OLD QUEEN'S LEAVING.

That the old queen does leave with the first swarm is indicated by several things: one is, eggs may often be found on the board the next morning; another, when the first swarm has left, and before any of these royal cells hatch, the bees may be driven out and no queen will be found, or you may drive out the bees at the end of three weeks, and the brood of workers will be about all hatched, the drone brood not quite as near. The combs may also contain some eggs, and perhaps some very young larvæ, that have been deposited by the young queen, which begins to lay usually sixteen or eighteen days after the first swarm. This shows a cessation of laying eggs for about two weeks. First swarms will have eggs in the cells as soon as they are made to hold them, which is often within 24 hours after being hived; occasionally a new piece of comb will fall down, and, if the cells are deep enough, they are almost certain to contain eggs. I could add other proof, but the attentive observer will discover it himself.

MR. WEEKS' THEORY NOT SATISFACTORY.

Mr. J. M. Weeks, in his work on bees, says, "Two causes and two only can be assigned why bees ever swarm: the first, the crowded state of the hive; the second, to avoid the battle of the queens." The first cause producing first swarms, the other second, third, &c. Mr. Colton's patent hive, it is said, can be made to swarm "at any time within two days," merely for want of room. By removing the six boxes attached to it, the bees are compelled to crowd into the main body of the hive, and swarm out in consequence. Now, if merely crowding the hive with bees is the only cause of first swarms, how is it that half or more of mine refused to swarm, when a great many, for want of room, were crowded outside for weeks, and great numbers maturing every day to crowd them still more? To me the reason is plain, that some of the before-mentioned requisites were wanting. Mr. Weeks further says, when the first swarm has left, "not a single queen, in any stage of minority, is left in the old hive; the bees, destitute of a queen, set about constructing several royal cells, take larvæ or eggs and put in them, and feed with royal jelly, and in a few days have a queen." Although I had not had much experience at the time of getting his work, I had some doubts, because I found that all hives that became full and began to run over, did not swarm, and some others swarmed before being quite full; it seemed as if something like a preparation beforehand was requisite. I knew of no means, for a long time, that would decide *positively*; when it occurred to me, if I examined the old stock immediately after the first swarm had left, I should find some preparations if there were any; a thing so simple and easy that I felt somewhat mortified not to have thought of it before. The first stock I looked at revealed the secret. I examined it the evening of the day that a swarm had left; I was gratified by finding two finished cells on the lower edges of the combs; other cells were in different stages of progression, from those containing an egg to the full developed larva. Several more hives showed the same result. I now got bold enough to examine some previous to swarming, as I have already explained.

MR. MINER NOT CORRECT.

Mr. T. B. Miner, in his work, has allowed the preparation of queen cells previous to swarming, but he has put off the time of the swarm issuing eight or nine days too long. That is, he has the young queen matured so that she commences piping first, which does not occur more than one time in fifty.

Now I think it more than probable that many readers will have some doubts in regard to my statements about this swarming matter. Yet I think I can give directions sufficiently particular that they may remove them themselves. They should bear in mind that they have no right to be *positive* on any subject without an investigation.

PARTICULAR DIRECTIONS FOR TESTING THE MATTER.

I will now give more minute directions for an examination. Full hives require a little more care than those containing fewer bees. Don't let the crowded state of the hive, even if some are outside, deter you from gratifying a laudable curiosity, (such hives are most likely to possess these cells.) Let the satisfaction of ascertaining a few facts for yourselves stimulate you to this exertion, the risk is not much; what I have done you may do. This is better than to rely on any man's "*ipse dixit*." I do it without any protection whatever for face or hands; but, if you have too much fear of stings, a veil to protect the face may be put on, but do without it, if you can find the courage, as you will want a good view. The best time is, when most of the bees are out at work near the middle of the day; but then the bees from the other hives are sometimes cross, and interfere. On that account I prefer morning or evening, although there are more bees to be smoked out of the way. If you are accustomed to smoking tobacco, you will find a pipe just the thing for making a smoke here; if not, vide a description of an apparatus in chap. 18th, p. 281. When you are ready to proceed, some smoke must be blown under the hive before you touch it; then raise the front side a few inches, and blow in some more; now carefully lift the hive from the stand, avoiding any jar, as this would arouse their anger; turn it bottom upwards; also, be careful all the time not to breathe among them. More smoke will now make them crowd among the combs out of your way while you examine. It is very common for the bees to set up a buzzing, and rush up the sides of the hive, but a little smoke will drive them back; get them out of the way as much as possible, and look on the edges of the combs for the queens' cells, where most of them are. If the hive is fully supplied with honey, they will be near the bottom, if not, farther up among the combs; in some hives they cannot be seen even where they exist. Yet they may be found in four out of five, by a thorough search. I have found nine within two inches of the bottom, some on the extreme ends of the comb. I would here give a caution about turning over hives with very new combs, before they are attached to the sides of the hive, as they are apt to bend over.

EMPTY HIVES TO BE READY.

We will now suppose that some of your stocks are ready to cast their swarms: we will also presume that your empty hives for the reception of swarms are ready before this period; to prepare a hive after the swarm has issued is bad management; negligence here argues negligence elsewhere; it is one of the premonitions of "bad luck."

BOTTOM-BOARDS FOR HIVING.

You will want also a number of bottom-boards, expressly for hiving; get a board a little larger than the bottom of the hive, nail strips across the ends on the under side to prevent warping; in the middle cut out a space five or six inches square, and cover with wire cloth. These are for your large swarms in very hot weather, to be used for four or five days. They are much safer than to raise the hive an inch or more for ventilation. They are also essential for many other occasions. I would not do without them, even if the expense was ten times what it is.

DESCRIPTION OF SWARM ISSUING.

When the day is fair and not too much wind, first swarms generally issue from ten o'clock till three; if you are on the lookout, the first outside indication of a swarm, will be an unusual number of bees around the entrance, from one to sixty minutes before they start. The utmost confusion seems to prevail, bees running about in every direction; the entrance apparently closed with the mass of bees, (perhaps one exception in twenty,) presently a column from the interior forces a passage to the open air; they come rushing out by hundreds, all vibrating their wings as they march out; and when a few inches from the entrance, rise in the air; some run up the side of the hive, others to the edge of the bottom-board. If you have seen the old queen come rushing out the first one, and the rest following her, as we are often told she does, you have seen what I never did in a first swarm! Second and third swarms conduct themselves quite differently. I have seen the old queen issue a few times, but not till half the swarm was out.

The bees when first rising from the hive, describe circles of but few feet, but as they recede, they spread over an area of several rods. Their movement are much slower than usual, in a few minutes thousands may be seen revolving in every possible direction! A swarm may be seen and heard, at a distance, where fifty hives, ordinarily at work, would not be noticed! When about out of the hive, or soon after, some branch of a tree or bush is usually selected on which to cluster. In less than half a minute after the spot is indicated, even when the bees are spread over an acre, they are gathered in the immediate vicinity, and all cluster in a body from five to ten minutes after leaving the hive. They should now be hived immediately, as they show impatience if left long, especially in the sun; also, if another stock should send out a swarm while they were hanging there, they would be quite sure to mix together.

MANNER OF HIVING CAN BE VARIED.

It makes but little difference what way they are put in the hive, providing they are all made to go in. Proceed as is most convenient; an old table or bench is very good to keep them out of the grass if there should happen to be any; if there is not much in the way, lay your bottom-board on the ground, make it level, set your hive on it, and raise one edge an inch or more to give the bees a chance to enter.

USUAL METHOD.

Cut off the branch with the bees, if it can be done as well as not, and shake it in front of the hive, a portion will discover it, and will at once commence a vibration of their wings; this, I suppose, is a call for the others. A knowledge of a new home being found seems to be communicated in this way, as it is kept up until all are in. A great many are apt to stop about the entrance, thereby nearly or quite closing it, and preventing others going in, when they will gather on the outside. You can expedite the matter with a stick or quill, by gently pushing them away; and another portion will enter. When gentle means will not induce them to go in, in a reasonable time, and they appear obstinate, a little water sprinkled on them will facilitate operations greatly, when nothing else will. (Be careful and not over-do the matter, by using too much water, they can be so wet as not to move at all.)

When they cluster on a branch that you do not wish to cut off, place your bottom-board as near as convenient; on it lay two sticks about an inch in diameter, of the same length: try the hive, and see that all is right; then turn it bottom up, directly under the main part of the cluster; if you have an assistant, let him jar the branch sufficiently to detach the bees; most of them will fall directly into the hive. If no assistant is at hand it is unnecessary to wait, (I have done it a hundred times without help); with the bottom of the hive strike the under side of the branch hard enough to dislodge them, then turn it on the board; the sticks will prevent the bottom crushing many bees.

WHEN OUT OF REACH.

I have gone up a ladder fifteen feet, got the bees in the hive in this way, and backed down without difficulty. After putting the hive in its place, sometimes a part will go back; in that case, a small branch full of leaves should be held directly under and close to them, and as many jarred on it as possible. Hold this still, and shake the other to prevent their clustering there; you will soon have them all collected, ready to bring down, and put by the hive. A handle basket or large tin pan may be taken up the ladder instead of the hive, when they can be readily emptied before it. But very few will fly out in coming down. If you succeed in getting nearly all the bees in the first effort, and but few are left, merely shaking the branch will be sufficient to prevent their holding fast, and will turn their attention to those below, where those which have already found a hive will be doing their best to call them. When the hive is first turned over, most of the bees fall on the board and rush out, but as soon as it is realized that a home is found, a buzzing commences inside; this quickly communicates the fact to those outside, which immediately turn about, facing the hive and hum in concert, while marching in.

Another plan may be adopted, even if fifteen feet high; when the branch is not too large, and there is not too much in the way below it. Have ready two or three light poles of suitable length; select such as have a branch at the upper end, large enough to hold a two-bushel basket. This is raised directly under the swarm; with another pole, the bees are all dislodged, and fall into the basket, and are quickly let down. Now, if you have got about all, throw a sheet over for a few moments, to prevent their escape. They soon become quiet, and may be hived without many going back to the branch, as they do, when attempting to hive them immediately.

I often have them begin to cluster near the ground, very conveniently for hiving. In such a case, I do not wait for all to collect, but as soon as such place is indicated, I get the board and hive ready. When a quart or so are gathered, shake them in a hive, and set it up; the swarm will now go to that, instead of the branch, especially if the latter is shaken a little. Where many stocks are kept, it is advisable to be as expeditious as possible. A swarm will thus hive itself much sooner than when it is allowed to cluster.

WHEN THEY CANNOT BE SHAKEN OFF.

Swarms will sometimes get in places where it is impossible to jar them off, or cut off a branch, such as the trunk of a tree, or a large limb near it. In which case place the hive near, as first directed; take a large tin dipper, a vessel most convenient for the purpose, and dip it full of bees; with one hand turn back the hive; with the other throw the bees into it; some of them will discover that a home is provided, and set up the call for the rest, (by the vibration of their wings), and the remainder may be emptied in front of the hive as you dip them off. I have known a few instances when the first dipper full all ran out, and joined the others without making the discovery that they were in a hive, but this is seldom the case. When you get the queen in, there is no trouble with the remainder, even if there are many left; as soon as they ascertain that the queen is no longer among them, it may be known by their uneasy movements, and they will soon leave, and join those in the hive; but if the queen is yet on the tree, and but a dozen with her, they will leave the hive and cluster again.

ALL SHOULD BE MADE TO ENTER.

In all cases be sure to get them all to enter; a cluster outside of it may contain the queen, unconscious of a home so near; and the probable consequence might be, she would leave for a miserable one in the woods.

SHOULD BE TAKEN TO THE STAND IMMEDIATELY.

When all are in, except a few that will be flying, let the hive down close to the board; take hold of this and carry it at once to the stand they are to occupy, and raise the front edge half an inch; let the back rest on the board; this will give them means to re-ascend, if they chance to drop, which large swarms often do in hot weather. If the bottom is an inch or more from the board when the bees fall, there is nothing to prevent their rushing out on every side—their means of getting up again are bad—if the queen comes out with the rush, there are some chances for their leaving.

PROTECTION FROM THE SUN NECESSARY.

Another thing is very important; *swarms should be protected from the sun for several days, in hot weather,* from nine o'clock till three or four; and then if the heat is very oppressive, and the bees cluster outside, sprinkle them with water and drive them in; and by wetting the hive occasionally, it will carry off a large portion of the heat, and make it much more comfortable.

CLUSTERING BUSHES.

If there are no large trees in the vicinity of your apiary, all the better, as there will then be no danger of your swarms lighting on them; but all bee-keepers are not so fortunate, myself being one of the number. In such a place it is necessary to provide something for them to cluster on; get some bushes six or eight feet high (hemlock is preferable); cut off the ends of the branches, except a few near the top: secure the whole with strings to prevent swaying in ordinary winds; make a hole in the earth deep enough to hold them, and large enough to be lifted out easily. The bees will be likely to cluster on some of these; they can then be raised out, and the bees hived without difficulty. A bunch of dry mullein tops tied together on the end of a pole, makes a very good place for clustering; it so nearly resembles a swarm that the bees themselves appear to be sometimes deceived. I have frequently known them leave a branch where they had begun to cluster, and settle on this when held near.

The motives for immediately removing the swarm to the stand are, that they are generally more convenient to watch in case they are disposed to leave; also many bees can be saved. All that leave the hive, mark the location the same as in spring; several hundreds will probably leave the first day; a few may leave several times; when removed at night, such will return to the stand of the previous day, and generally are lost; whereas, if they are carried at once to a permanent stand, this loss is avoided.

Those that are left flying at the time, return to the old stock, which those that return from the swarm the next day will not always do. The time for moving them now is no more than at another. It is unnecessary to object, and say, that "it will take too long to wait for the bees to get in;" this will not do. I shall insist on your getting all the bees to enter before leaving any way. I consider this an essential feature in the management. I will not say that my directions will *always* prevent their going to the woods, but this I do say, that out of the hundreds that I have hived, not one has ever left. It is possible proper management has had no influence in my success, yet something like an opinion of this kind has been indulged for a long time.

HOW SWARMS ARE GENERALLY MANAGED THAT LEAVE FOR THE WOODS.

Some of my neighboring bee-keepers lose a quarter or half of their swarms by flight, and how do they manage? When the word is given out, "Bees swarming," a tin-horn, tin-pan, bells, or anything to make a "horrible din," is seized upon in the hurry of the moment, and as much noise made as possible, to *make* them cluster; (which they naturally would do without the music, at least all mine have. This probably gave rise to the opinion of one old lady, who *knew* "drumming on a pan did good, for she had tried it.") Very often a hive is to be constructed, or an old one unfit to use any way, needs some sticks across, or something to take time. When the hive is obtained, it must be washed with something nice to make the bees like it; a little honey must be daubed on the inside; sugar and water, molasses and water, salt and water, or salt and water rubbed on with hickory leaves, "is the best thing in the world;" several other things are just as good, and some are better. Even whisky, that bane of man, has been offered them as a bribe to stay, and sometimes they are persuaded and go to work.

NOTHING BUT BEES NEEDED IN A HIVE.

Now I cannot say positively that these things do harm, yet I am quite sure they do no good, as nothing but bees is needed in a hive. Is it reasonable to suppose they are fond of all the "knick-knacks" given them? I have never used any, and could not possibly have done better. I am careful to have the hive sweet and clean, and not too smooth inside; an old hive that has been used before is scalded and scraped.

But to the manner they get the bees in, after the hive is ready. A table is set out, and a cloth spread on it; sticks are put on to raise the hive an inch or more: if they succeed in getting the swarm even on the outside of the hive it is left; if they go in, it is well; if they go off, why hope for "better luck next time." The hive is left unsheltered in the hot sun and when there is no wind, the heat is soon insupportable, or at least very oppressive; the bees hang in loose strings, instead of a compact body, as when kept cool; they are very apt to fall, and when they do, will rush out from every side: if the queen chances to drop with them, they *may* "step out." Two thirds of all the bees that go to the woods are managed in this, or a similar manner, and may it not be said, they are fairly driven off?

SELDOM GO OFF WITHOUT CLUSTERING.

Perhaps one swarm in three hundred will depart for the woods without first clustering. I have had three times that number, not one of which has ever left me thus. Yet I have evidence not to be disputed that some will do it. Three instances have occurred near me that satisfied me of the fact. Two were lost, the other was followed to a tree, half a mile off; I assisted in cutting the tree, and hiving them. The cavity where they entered was very small, and contained old comb, made by a swarm a year or two previous, which had probably starved, as there was too little room for storing sufficient honey for winter. This swarm, when hived and carried home, remained perfectly contented.

DO SWARMS CHOOSE A LOCATION BEFORE SWARMING?

The inquiry is often made, Do all swarms have a place looked out before leaving the parent stock? The answer to this must ever be guess-work. I could offer some circumstances indicating the affirmative very strongly, and as much for the negative; and will let it pass at that. Yet I think if bees are properly cared for, that ninety-nine swarms in a hundred will prefer a good clean hive to a rotten tree in the woods.

MEANS OF ARRESTING A SWARM.

I have had three swarms that were exceptions to general rules, giving me some trouble by swarming out after being hived; the third and fourth time they left, I threw water among them, causing quite a shower; when my pail-full was out, I used earth; they went but a short distance, and clustered in the usual way. Now were these bees intending to leave, and had their designs frustrated by the water and earth? I am not quite as sure as the old lady, who *knew* that "drumming on a tin-pan did good," but I am inclined to think it had some effect. I have heard of several instances where swarms were apparently stopped, by having earth thrown among them, while passing over a field where men were at work. We know they dislike being wet, as we see them hastening home on the approach of a shower; or we can at any time drive them in the hive by sprinkling them with water. Throwing water in the swarm is a kind of imitation shower, and earth is something like it. Whether useful or not, these swarms leaving the hive was rather suspicious, and I should try it again under similar circumstances.

SOME COMPULSION.

After getting them in the hive for the fourth time, I resolved not to be baffled or have much more such trouble, and perhaps go to the woods at last, thereby setting a bad example. I put under the hive the wire-cloth bottom-board, opened two or three holes on the top, and covered these also with wire-cloth, (this was to let the air circulate); a quantity of honey and water was given them and they were then carried to the cellar, and kept prisoners four days, except half an hour before sunset; when too late to leave for a journey, I set them out to provide a few necessaries, and then returned them to the cellar. In four days, when *honey enough* is given them, a good swarm will half fill an ordinary hive with combs. Some of the first eggs deposited will be about hatching into larvæ, all of which would seem like too much to leave. I now set them out, and gave them liberty; shading the hive, &c., as before directed. They all proved faithful and industrious, prospering like others. If their design was for a distant location, they put a good face on the matter in the end.

HOW FAR WILL THEY GO IN SEARCH OF A HOME?

How far they will travel in search of a home, is also uncertain. I have heard of their going seven miles, but could not learn how the fact was proved. I have no experience of my own in this matter, but will relate a circumstance that happened near me a few years since. A neighbor was ploughing, when a swarm passed over him; being near the earth, he "pelted them heartily" with the loose dirt he had ploughed up, which seemed to bring them up, or rather down, as they clustered on a very low bush; they were hived, and gave no further trouble. A man living some three miles from this neighbor, on that day hived a swarm about eleven o'clock, and left them to warm up in the sun as described a page or two back; about three o'clock their stock of patience was probably exhausted, when they resolved to seek a better shelter. They put off in a great hurry, not even waiting to thank their owner for the spread on his table, and the sweet-scented "yarbs" and good things with which he had rubbed their hive. They gave him no notice whatever of their intention to "quit," until they were moving! With all their goods ready packed, they were soon under way, accompanied by their owner with music; but whether they marched with martial precision, keeping time, is uncertain. In this case the bees took the lead; the man with his tin-pan music kept the rear, and was soon at a respectful distance. They were either not in a mood, just then, to be charmed by melodious sounds, or their business was too urgent to allow them to stop and listen! Their means of locomotion being superior to his, he gave up in despair, out of breath, after following about a mile. Another person, about the same time in the day, saw a swarm moving in the same direction of the first; he also followed them till compelled to yield to their greater travelling facilities. A third discovered their flight and attempted a race, but like the others soon came out behind. The before-mentioned neighbor saw them, and thought of the fresh earth that he had ploughed up, which he threw among them till they stopped. How much farther they would have gone, if any, would be guessing. That it was the same swarm that started three miles away, appears almost certain; the direction was the same as seen by all, until they were stopped; the time in the day also exactly corresponded.

We will now return to the issuing of the swarms. There will be some emergencies to provide for, and some exceptions to notice.

TWO OR MORE SWARMS LIABLE TO UNITE.

If we expect to keep many stocks, the chances are that two or more may issue at one time; and when they do, they nearly always cluster together (I once knew an instance where only three stocks were kept; they all swarmed and clustered together). It is plain that the greater the number of stocks, the more such chances are multiplied.

DISADVANTAGE.

One first swarm, if of the usual size, will contain bees enough for profit, yet two such will work together without quarrelling, and will store about one-third more than either would alone; that is, if each single swarm would get 50 lbs., the two together would not get over 70 lbs., perhaps less. Here, then, is a loss of 30 lbs., besides one of the swarms is about lost for another year; because such double swarms are not generally any better the next spring as a stock, and often not as good as a single one. You will therefore see the advantage of keeping the first swarms separate.

CAN OFTEN BE PREVENTED.

"Prevention is better than cure." We can, if we keep a good lookout, often prevent more than one issuing at a time. This depends on our knowledge of indications, in a great measure. I have said that before starting to fly off, they were about the entrance in great numbers; there may be one exception in twenty, where the first indications will be a column of bees rushing from the hive. To take this matter a little farther from the surface, we will take a peep at the interior; that is, if our hives contain glass boxes, such as have been recommended. It is an advantage to know which are about to cast their swarms, as long beforehand as possible.

INDICATIONS OF SWARMING INSIDE THE HIVE.

These glass boxes are usually filled with bees; before leaving they may be seen in commotion, long before any unusual stir is visible outside, sometimes for near an hour. The same may be noticed in a glass hive. Now, in good weather, when we have reason to expect many swarms, it is our duty to watch closely, especially when the weather has been unfavorable for several days previous. A number of stocks may have finished their queen-cells during the bad weather, and be ready to come out within the first hour of sunshine that occurs in the middle of the day. We must expect some such occurrences, and in large apiaries there is apt to be trouble, unless you take some precautions. If you have taken no care (which but few will), by previous examinations, to know which are ready, as soon as one has started or commenced flying, look at all the rest that are in condition to swarm; or, what is much better, look before any have started. Even if nothing unusual is seen about the entrance, raise the cover to the boxes. If the bees in them are all quiet as usual, no swarm need be immediately apprehended, and you will probably have time to hive one or two first.

PREVENTING A SWARM ISSUING FOR A TIME.

But should you discover the bees running to and fro in great commotion, although there may be but few about the entrance, you should lose no time in sprinkling those outside with water from a watering-pot, or other means. They will immediately enter the hive to avoid the supposed shower. In half an hour they will be ready to start again, in which time the others may be secured. I have had, in one apiary, twelve hives all ready in one day, and did actually swarm; several of which would have started at once, had they not been kept back with water, allowing only one at a time, thus keeping them separate. They had been kept back by the clouds, which broke away about noon.

TO PREVENT SWARMS UNITING WITH THOSE ALREADY HIVED.

When any of the subsequent swarms were disposed to unite with those already hived, a sheet was thrown over to keep them out. I had four so covered at once. An assistant, in such cases, is very important; one can watch symptoms and keep them back, while the other hives the swarms.

Occasionally, when ready for a swarm and waiting for one to start, two may do so at once. Whenever a part have got on the wing, I never succeeded in stopping the issue: consequently I have found it useless trying to drive or coax them back in such cases. To succeed, the means must be used in season, before any of the swarm leaves.

WHEN TWO HAVE UNITED, THE METHOD OF SEPARATING.

Two or more swarms will cluster together, and not quarrel, if put in one hive; I have already told you the disadvantages. Unless business is very urgent, your time cannot be better employed than in dividing them. First, it is necessary to provide a good stock of patience, as it may be a short job, or it may be a long one. Get two empty hives, and divide the bees as nearly equal as possible. It is generally the best way to spread a sheet on the ground, and shake the bees in the centre, and set the hives each side of the mass, their edges raised to allow the bees to enter; if too many are disposed to enter one hive, set it farther off. If they cluster in a situation where they cannot be got to the earth in a body, they must be dipped off as before directed, but, in this case, putting a dipper full in each hive alternately, until all are in. They should be made to hurry some in going in; keep the entrance clear, and stir them up often; or sprinkle a very little water on them, as they should not be allowed to stop their humming until all are in. We have one chance in two of getting a queen in each. The two hives should now be placed twenty feet apart; if there is a queen in each, the bees in both will remain quiet, and the work is done; but if not, the bees in the one destitute will soon manifest it by running about in all directions, and, when the queen cannot be found, will leave for the other hive, where there are probably two, a few going at a time. Now there are two or three methods of separating these queens; one is, to empty the bees out and proceed as before, a kind of chance game, that may succeed at the next trial, and may have to be repeated. Another way is, that, as soon as it is ascertained which is without a queen, before many bees leave, spread down a sheet; set this hive on it, and tie the corners over the top to secure the bees for the present, turn the hive on its side for the present to give them air; or it may be let down on a wire cloth bottom-board and the hole in the side stopped, and this would be less likely to smother the bees, if it could be secured to the bottom, and have the hive lie on its side; when this division is secured, get another hive, and jar out those with the queens; let them enter as before, and then set them apart, &c., watching the result; if the queens are not yet separate, it will be known by the same appearances. The process must be continued till separate, or the number with the queens may be easily looked over, and one of them found; indeed, a sharp lookout should be kept up from the beginning, and the queens caught, if possible.

NO DANGER OF A STING BY THE QUEEN.

No danger of her sting need be apprehended, as she will not demean herself to use that for a common foe; she must have a *royal* antagonist. When successful in obtaining one, it is sufficient; put her in a tumbler or some safe place; then put your bees in two hives, place them as directed, and you will soon learn where your queen is needed. After all is done, the two hives should not be nearer than twenty feet, at least the first day; perhaps forty would be still better. When two swarms are mixed, and then separated, it is evident that a portion of each swarm must be in both hives. A queen in each must of course be a stranger to at least a part of the bees; these might, if their own mother was too near, discover her, and leave the stranger for an old acquaintance, and, in the act of going, call or attract the whole with them, including the queen. I have known a few instances of the kind.

SOME PRECAUTIONS IN HIVING TWO SWARMS TOGETHER.

If you are disposed to separate them, but are afraid to work among them to this extent in the middle of the day, or if there is danger of more issuing, to mix with them, and add to your perplexity, of which you already have enough, then you can hive them as a single swarm; but, instead of a bottom-board, invert an empty hive and set the one with the swarm on this, and insert a wedge between them, for ventilation. As many bees are liable to drop down, in this case the lower hive will catch them, and there is less danger of leaving. Let them remain till near sunset, when another course may be taken to find a queen, though by that time one is sometimes killed; yet it is well to know the fact. Take them to some place out of the sun, as a less number will fly during the operation.

HOW TO FIND QUEEN, WHEN TWO STRANGERS ARE TOGETHER.

First, look into the lower hive for a dead queen, and, if none is found there, look thoroughly, as far as possible, for a little compact cluster of bees, the size of a hen's egg, that may be rolled about without separating. Secure this cluster in a tumbler; it is quite sure one of the queens is a prisoner in the middle;[16] should two be seen, get both. Then divide the bees, and give the one destitute, a queen; or, if you have two, one to each, as the case may be. It would be well first to see if the queen was alive, by removing the bees from about her. But should you find nothing of the kind, spread a sheet on the ground, shake the bees on one end of it, and set the hive on the other; they will immediately begin a march for the hive. You may now see the cluster, and may not; but they will spread out in marching, and give a good chance to see her majesty, when a tumbler is the most convenient thing to set over her. No matter if a few bees are shut up with her, there is no risk, then, in your eagerness to get the queen, of taking hold of a worker or two. A piece of window-glass can be slipped under, and you have her safe, and by this time you will know what is to be done next. This operation could not well be done in the middle of the day, or in the sun, as too many bees would be flying, and greatly interfere.

Should you fail in finding a queen, and cannot succeed in making a division in consequence, or should you resolve, from want of time, patience or energy, to let them remain together in the beginning, it is unnecessary to get a hive any larger than usual for two swarms; they will certainly find room by cold weather: if more than two, they *should* be divided by all means; it will be a disadvantage for another year. For the first four days, when two large swarms are together, it is necessary to keep an inverted hive under them, but much longer it would not do, as they might extend their combs into the lower hive.

BOXES FOR DOUBLE SWARMS IMMEDIATELY.

It should then be taken out, and boxes immediately put on, which should be changed for empty ones, as fast as they are filled. Yet, this extra honey is not quite as much advantage as increase of stocks; when that is an object, I will recommend another disposition.

RETURNING A PART TO THE OLD STOCK.

Separate one-third or more of the two swarms, being sure there is no queen with this part, (by the test given of setting them at a distance) and then return them to one of the old stocks; they will immediately enter without contention, and issue again in about nine days, or as soon as a young queen is matured to go with them. There may be an exception to this, of one in twenty. I would have recommended this course in all cases of the kind, but there will be a loss of time for the bees in the old stock; because they are apt to be rather idle, even when they might labor in the boxes; and here there is a loss of some eight or ten days. The collections of a good swarm may be estimated at least one pound a day, (often two or three.) A swarm that just fills the hive, would make at least ten pounds box-honey, if it could have been located ten days earlier. Still another method may be adopted when you have a very small swarm, one that is not likely to fill the hive, and has not been hived more than two or three days. A third of your two swarms may be put in with that; taking care, as before, not to let your only queen go with them.

METHOD OF UNITING.

The manner of doing it is very simple; get them in a hive as before directed, and jar them out in front of the one you wish them to enter, or invert it, setting the other over, and let them run up.

WHEN CARE IS NECESSARY.

Except on the day of swarming, care is necessary not to introduce a small number with a large swarm; they are liable to be destroyed. The danger is much greater than to put together about an equal number, or a large number put in with a few. The day that swarms issue, they will generally mix peaceably, but in proportion as time intervenes between the issues, so will be the liability to quarrel. Yet, I have united two families of about equal numbers in the fall and spring, and, with a few exceptions, have had no difficulty.

SWARM-CATCHER.

There is another method of keeping swarms separate, contrived and used by a Mr. Loucks, of Herkimer Co., N.Y. He calls it a swarm-catcher; he has a half dozen of them, and says he would not do without for one season, for fifty dollars, as he has a large apiary. I made one as near as I could from seeing his, without taking the exact measure. I got out four light posts four and half feet long, one inch square; then twelve pieces of one-quarter inch stuff, four inches wide; the four for the top twelve inches long, for the bottom two were fourteen inches long, and two were twenty. These were thoroughly nailed on the ends of the posts, making it into an upright frame, the other four pieces were nailed around the middle, which made the frame firmer. I made a frame for the top, of four pieces, each an inch and a half in width, and half inch thick, halved at the ends and nailed together, and fastened by hinges to one side of the top, and a catch to hold it shut. The whole was now covered with very thin cloth to admit the light, but not so open as to let the bees through, (Mr. Loucks used cloth made for cheese-strainers.) I now had a covered frame four and half feet high, 12 inches square at the top, at the bottom 14 by 20, with a door or lid at the top, to let out the bees. On each side of the bottom I tacked a piece of common muslin, near a yard in length. When a swarm is ready to issue, the bottom of this frame is set up before the hive, one edge of the bottom rests on the bottom-board, the other against the side of the hive; the top sets off from the hive at an angle of about 45 degrees, under which a brace is set to hold it. The muslin at the bottom is to wrap around the hive at the side to prevent the escape of the bees. The swarm rushes into this without any hesitation.

When done coming out, the muslin at the bottom is drawn over it, and the frame is set in an upright position, and allowed to stand a few minutes for the bees to get quiet in the top. It is now to be laid on its side, the door opened, and the bees hived. In the few trials that I have given it, I succeeded without difficulty. But I would remark, that stocks from which swarms are caught in this way, must not be raised at the back side, as a part of the swarm would issue there, and not get into the net. Mr. Loucks had his hive directly on the board; and he told me he kept them so through the season: the only places of entrance was a sprout out of the bottom of the front side, about three inches wide by half inch deep, and a hole in the side a few inches up. You will thus perceive that stocks from which swarms are hived in this way must be prepared for it previously. Also, it will be no use to such bee-keepers as depend on seeing their swarms in the air. It will be beneficial only in large apiaries, where several swarms are liable to issue at once; the swarming indications well understood, and the apiarian on the lookout.

SWARMS SOMETIMES RETURN.

Occasionally a swarm will issue, and in a few minutes return to the old stock. Mr. Miner gives a cause for this, very ingenious, and romantic, but unfortunately there are but few facts to sustain this hypothesis, (at least I have not discovered them.) There are other causes that appear to me more reasonable; the most common is the inability of the old queen to fly, on account of her burden of eggs, old age, or something else. I have sometimes, after the swarm had returned found the queen near the stock, and put her back, and the next day she would come out again, and fly without difficulty, (perhaps she had discharged some of her eggs.)

Their returning is more frequent in windy weather, or when the sun is partially obscured by clouds. About three-fourths of them will not re-issue until a young queen is matured, eight or ten days afterwards; and a few, not at all. But when the queen returns with the swarm, they usually come out again the next day, or day after, and some not till the third or fourth. I have known two instances where they issued again the same day.

REPETITION PREVENTED.

Sometimes a swarm will issue and return three or four days in succession, but this I generally remedy, as it is often owing to some inability of the queen, and she may be frequently found while the swarm is leaving outside the hive, unable to fly. In such cases it is only necessary to have a tumbler ready, and watch for her; and as soon as she appears, secure her, get the empty hive for the swarm, a sheet, and put down a bottom-board a few feet from the stock.

The swarm is sure to come back; the first bees that alight on the hive will set up the call; as soon as this is perceived, lose no time in setting the old stock on the board, and throwing the sheet over it to keep out the bees. Put the new one in its place on the stand, and the queen in it; in a few minutes the swarm will be in the *new* hive, when it can be removed, and the old one replaced. This I have done many times. But should the swarm begin to cluster in a convenient place, when you have so caught the queen, by being expeditious she may be put with the swarm, before they have missed her and may be hived in the usual way.

LIABILITY TO ENTER WRONG STOCKS.

In all cases, whether you set a new hive in place of the old one or not, whenever a swarm returns, if other stocks stand close on each side, they are quite sure to receive a portion of the bees—probably a few hundreds; these are certain to be massacred. To prevent which, it is necessary to throw sheets over them until the swarm has gathered on their own hive. This is another reason for plenty of room between stocks. Should no queen be discovered during their issue, or return, she should be sought for in the vicinity of the hive, and put back if found, and the swarm will be likely to issue several days earlier, than to wait for a young queen.

When the old queen is actually lost, and the bees have returned to wait for a young one, it is often ready to leave one or two days short of the time required for second swarms. Whether a greater number of bees in the old stock creating more animal heat, matures the chrysalis in less time than a stock thinned by casting a swarm, or some other cause, I cannot say. I mention it because I have known it to occur frequently, but not invariably. A swarm flying, unaccompanied by a queen, is scattered more than usual.

FIRST ISSUES GENERALLY CHOOSE FAIR WEATHER.

First swarms are commonly more particular as to weather than after swarms. They have several days from which to choose, after these royal cells are ready, and before the queens are matured; and they usually take a fair one. But here again are exceptions. I once had two first swarms issue in a wind that kept every branch of tree and bush in agitation to such a degree that it was impossible to find any such place to cluster. I expected their return to the old hive; but here were more exceptions. After repeating a fruitless attempt at the branches, they gave it up, and came down amongst the grass on "terra firma." This occurred after several days of rainy weather. The next day being pleasant, twelve issued; almost proving that the wind the preceding day kept back a part. I also knew one to issue in a shower, that beat many of them to the ground before they could cluster. In this case the shower was sudden, the sun shone almost up to the time it began to rain. About this time the swarm started when it seemed they were unwilling to turn about.

AFTER SWARMS.

After swarms are second and third issues (or all after the first) from a stock; and quite a different affair from the first, as also are some first swarms, when the old queen has been lost, being led out by young queens.

THEIR SIZE.

Second swarms are usually half as large as the first, the third half as large as the second, the fourth still less; with some variations. I give general features, noticing only the exceptions that occur most frequently; others sometimes happen, but so seldom that mentioning them is deemed unnecessary.

TIME AFTER THE FIRST.

Whenever the first swarm in a prosperous season *was not kept back by foul weather*, the first of the young queens in the old stock is ready to emerge in about eight days. We will suppose the first swarm issued on Sunday; a week from the next Tuesday will be usually as soon as the second one need be expected.

PIPING OF THE QUEEN.

On the Monday evening previous, or on Tuesday morning, by putting your ear close to the hive, and listening attentively five minutes, you will hear a distinct piping noise, like the word *peep, peep*, uttered several times in succession, and then an interval of silence; two or more may be often heard at the same time; that of one will be shrill and fine, of another hoarse, short and quick. This piping is easily heard by *any* one not actually deaf, and not the least danger of its being taken for any humming; in fact, it is not to be mistaken for anything else *but piping*, even when you hear it for the first time. These notes can probably never be heard except when the hive contains a plurality of queens.

MAY ALWAYS BE HEARD BEFORE AND AFTER SWARM.

I *never failed to hear it*, previous to a second swarm, or any after the first, whenever I listened; and whenever I have listened and not heard it at the proper time, I never knew a second swarm to issue!

TIME OF CONTINUANCE VARIES.

The time of commencing will be later than this rule in some stocks, if the weather is cool, or not many bees left; it may be ten or twelve days. I once found it fourteen before I heard it. Also the swarm may not issue in two or three days after you hear it. The longer the swarm delays, the louder will be the piping; I have heard it distinctly twenty feet, by listening attentively when I knew one was thus engaged; but at first it is rather faint. By putting your ear against the hive it may be heard even in the middle of the day, or at any time before issuing. The length of time it may be heard beforehand seems to be governed again by the yield of honey; when abundant it is common for them to issue the next day; but when somewhat scarce, they will be much longer—very often three or four days. In these cases third swarms seldom occur.

TIME BETWEEN SECOND AND THIRD ISSUES.

Piping for third swarms (when they issue) may usually be heard the evening after the second has left, though one day commonly intervenes between their issues.

Here my experience is at variance with many writers, who give several days between the second and third. I do not recollect an instance of more than three days between, but many in less, several the next, and one the same day of the second! I had an instance of a swarm losing its queen (the old one) on its first sally, and returned to wait for the young ones; when they were ready, an uncommon number of bees were present; three swarms issued in three days! On the fourth, another came out and returned; the fifth day it left; making four regular swarms in five days. On the eighth, the fifth swarm left! Although I never had five swarms from a stock before, yet I expected this, from the fact of hearing the piping on the next evening after the fourth one had left. The piping had continued in this hive from the evening previous to the first swarm till the last one had left.

NOT ALWAYS TO BE DEPENDED UPON.

One stock in fifteen may commence piping, yet send out no swarm. The bees will change their minds about coming out, and kill their queens, or allow the eldest one of them to destroy the others, or some other way, as they do not always swarm in such circumstances. But when the piping continues over twenty-four hours, I never knew *but one failure*! I have known a few (two or three) to commence this piping, while I supposed the old queen was yet present, and had not left the hive, on account of bad weather, but a swarm issued soon after. Also, three instances where I supposed the old queen lost, from some other cause than leading out a swarm, and the stock reared some young ones to supply her place. It occurred in or near the swarming season, and one or two issues was the consequence. One case was three weeks in advance of the season, and the swarm was about half the usual size. When a swarm has been out, and returned at the last of the swarming season, it is much more probable to re-issue, than if it depended on an old queen for a leader, that had not been out. Such will sometimes be a week or ten days later than others. Once I had the first swarm kept back by wet weather, and the second came out on the fifth day after; several other instances on the seventh and eighth; and one as late as the sixteenth, after the first.

A RULE FOR THE TIME OF THESE ISSUES.

This may be put down as a rule, that all after swarms *must* be out by the eighteenth day from the first. I never found an exception, unless the following may be considered so: When a swarm left the middle of May, and another the first of July, seven weeks after, but two cases of this kind have come up, and these I consider rather in the light of first swarms, as they leave under the same circumstances, leaving the combs in the old stock filled with brood, queen-cells finished, &c. A stock may cast swarms in June, and a buckwheat swarm in August, on the same principle.

WHEN IT IS USELESS TO EXPECT MORE SWARMS.

Therefore, bee-keepers having but few stocks, will find it unnecessary to watch their bees when the last of the first swarms came out sixteen or eighteen days before. Much trouble may be thus saved by understanding this matter. During my early days in beekeeping, I wished for the greatest possible increase of stocks. I had some that had cast the first swarm, and soon after clustered out again. I vainly watched them for weeks and months, expecting another swarm. But had I understood the *modus operandi*, as the reader may now understand it, I should have been through with all my anxiety, as well as watching, in a fortnight. As it was, it lasted two months. I found no one to give me any light on this subject, or even tell me when the swarming season was over, and I came very near watching all summer!

PLURALITY OF QUEENS DESTROYED.

When the bees, queens, or all together, decide that no more swarms are to issue, the plurality of queens is destroyed, and but one is left. It is probable that the oldest and strongest queen dispatches the others, generally while in the cells.

I once had some artificial queens reared, as an experiment, from common eggs, on the top of a hive, in a small glass box, where there was room for but one comb, which allowed me to see all particulars.

THE MANNER.

After the first queen was matured, and had left her cell, I caught her within six hours, taking advantage of her younger sisters, which were yet sealed up, and of course could offer no resistance. She first made an opening that would allow her to reach the abdomen of her competitor (probably this is the most vulnerable). As soon as this was sufficiently large to admit her body, she thrust it in, inflicting the fatal sting. This was then left for another, that soon shared the same fate. If quick and spiteful movements are any indications of hatred, it was manifested here very plainly. The bees enlarged the orifice and dragged out the now dead queens.

Now, if I should say that all queens were dispatched in this way, merely because I witnessed it in this case, it would be carrying out the principle I am endeavoring to avoid: that is, judging all cases from one or two solitary facts. As it is, it is somewhat confirmatory of what some others have said. I will suppose, then, until further evidence contradicts it, that the first perfect queen leaving her cell, makes it her business to destroy all rivals in their cradle, as soon as it is decided that no more swarms shall issue. By keeping grass, weeds, &c., away from about the stock, these dead queens, as they are brought out, may be frequently found. Such as are removed during the night may be often found on the floor-board in the morning. I have found a dozen by one stock. Should the stock send out but one swarm, they may be found about the time, or a little before you would listen for the piping. But should after swarms come out, they will, or may be found the next morning after it is decided that no more are to issue. It is very seldom that all the queens reared are needed. They make it a rule, as far as they have control, to go on safe principles, by having a little more than just enough. When several such bodies are thrown out, and no piping is heard, no further swarming need be expected. But should you hear the piping a day or two after finding a dead queen, you may yet look for the swarm.

THEORY DOUBTED.

It is stated that when the bees decide an after swarm shall issue, the first queen matured is not allowed to leave her cell, but kept a prisoner there, and fed until wanted to go forth with the swarm. This may be true in some cases (though not satisfactorily proved), but I am quite sure it is not in all.

When she is confined to her cell, how does she ascertain the presence of others? By leaving the cell, this knowledge is easily obtained. Huber says she does, and is "enraged at the existence of others, and endeavors to destroy them while yet in the cell, which the workers will not allow; this is so irritating to her majesty that she utters this peculiar sound." Also second and third swarms may contain several queens, frequently two, three, and four; even six at one time come out. If these had to bite their way out, after the workers had decided it was time to start (for it *must be they* decide it when the queens are shut up), they would hardly be in season.

AFTER SWARMS DIFFERENT IN APPEARANCE FROM THE FIRST WHEN ABOUT TO ISSUE.

Another thing, when after swarms start, the appearance about the entrance is altogether different from first ones, unless there is an unusual number of bees. I have said that for a little time beforehand, that such were in an apparent tumult, &c. But after swarms seldom give any such notice. One or more of the young queens may sometimes be seen to run out, and back, several times in a few minutes, in a perfect frenzy; sometimes fly a short distance, and return before the swarm will get started (which she could not do if confined). The workers seem more reluctant about leaving than in first swarms, when a mother instead of a sister is leader. Even after the swarm is in motion, she may return and enter the hive a moment. No doubt she finds it necessary to animate or induce as many as possible to leave with her. A person watching the issue of a second swarm under these circumstances, for the first time, and finding the queen leaving first, would very likely *guess* all must be alike. Perhaps the next one would be different; the first thing seen might be the swarm leaving, and no queen discovered at all. But to return to the imprisonment of the queens. I have one other fact in objection. I once saw a queen running about in a glass hive, while they were piping for a second swarm. She was near the glass, appeared agitated, stopping occasionally to vibrate her wings, which was simultaneous with the piping, and seemed to make it. The workers appeared to take but little notice of her. The next day the swarm left. Here was one instance, at least, of her not being confined till the time of leaving, making an exception, if not a rule. Let this matter be as it may, I admit it makes but little difference to the practical apiarian, either way; but to the reader whose interest is the natural history of the bee, the truth is important.

TIME OF DAY, WEATHER, ETC.

These after swarms are not very particular about the weather; heavy winds, a few clouds, and sometimes a slight sprinkling of rain, will not *always* deter them. Neither are they very precise about the time of day. I have known them in a warm morning to issue before seven o'clock, and after five P.M. These things should be understood; because, when after swarms are expected (of which the piping will give warning), it is necessary to watch them in weather, and at times when first ones would not venture to leave.

SWARMS NECESSARY TO BE SEEN.

It is essential that you see them, that you may know where they cluster, otherwise it might be difficult to find them. They are apt to go farther from the parent stock than others; sometimes fifty rods, and then settle in two places, perhaps that distance apart, in some high or inconvenient place to get at. (Let me not be misunderstood: I do not say they all do so, or even the majority; but I wish to say that a greater portion of these swarms do so than of the first.) If they cluster in two places, a queen may be in each, and they will remain, and when you have hived one part you may think you have all. If one cluster is without a queen, they will join the other if near; but when distant, will be very likely to return to the old stock soon, unless put together. I had a swarm light in two places, in exactly opposite directions from the stock. In one, a good swarm had clustered; in the other, some less than a pint. The small part had one or more queens, the other none. It was perceived at once by their movements. Now, if we provide a hive for a swarm, and get a few to set up the call or buzzing, they will not leave till that is stopped. There is generally no difficulty to start it. The surest way is to jar a portion or all directly into the hive. It takes a few minutes to get composed, and miss the queen. In my case I got them in the hive, and before they missed the queen, carried them to the small cluster, which I got in a dipper and emptied in front of the hive; they entered, and all were peaceable. You will therefore see the necessity of watching such swarms, to see if there is no separation, if nothing else.

RETURNING AFTER SWARMS TO THE OLD STOCK.

Much has been said about returning all after swarms to the old stock; the advantages of which will depend on the time of issuing; whether late or early, the yield of honey, etc. It would be unusual to have many after swarms without a liberal yield of honey, for the time being; but to tell of its continuance is the question to be answered. Second, and even third swarms, if early in the season, and the honey continues plentiful, may be hived, and these, together with the old stock, will prosper. Here the apiarian needs a little judgment and experience to guide him.

WHEN THEY SHOULD BE RETURNED.

It is always best, if possible, to have good strong families. When after swarms are late it is safest to return them, as the old stock will need them to replenish the hive, and prepare for winter. Also a less number of worms will infest it, when well provided with bees; and the chances of box honey are greater.

METHOD OF DOING IT.

But the process of returning such requires some little patience and perseverance. I have said there may be a dozen young queens in the old stock. Now suppose one, two, or more leave with the swarm, and you return the whole together, there is nothing to prevent their leading out the swarm again the next day. Therefore it is policy to keep the queens back. The least trouble is to hive in the usual way, and let them stand till the next morning. It will save you the trouble of looking for more than one, if there should be more, for all but that are destroyed by that time. There is a chance, also, for the old stock to decide that no more should issue, and allow all but one to be slain there. When this is the case, and you find the one with the swarm, you will have no further trouble by their re-issuing. They should be returned as soon as the next morning, otherwise they might not agree, even when put in the old home. To return them, and find a queen easily, get a wide board a few feet long; let one end rest on the ground, the other near the entrance, that they may enter the hive without flying; then shake out the swarm on the lower end of the board; but few will fly, but soon commence running up towards the hive; the first one that discovers the entrance will set up the call for the others. If they do not discover it, which is the case sometimes, scatter some of them near it, and they will soon commence marching up, when you should look out for, and secure the queen, as they spread and give a good chance. By applying your ear to the hive, the piping will tell you if they are to issue again. It is evident, if you follow these directions, that the swarm cannot issue many times before their stock of royalty will be exhausted; and when but one queen remains the piping will cease, and no further trouble will be had. To prevent these after swarms, some writers recommend turning over the hive and cutting out all the royal cells but one. This I have found impracticable with a great many stocks. Some of the cells are too near the top to be seen, consequently this cannot always be depended upon. As for a rule about returning, it is somewhat difficult to give one. If I should say, return all such as issue after the 20th of June, the variation in the season might be two or three weeks, even in the same latitude; i.e., the course of flowers that had bloomed by that date in one season might, another year, require two weeks more to bring out. Also, the 20th of June, in latitude of New York City, is as late as the 4th of July in many places further north. I once had a second swarm on the 11th of July, that wintered well, having nearly filled the hive. Yet, in some seasons, the first swarms, of the last of June, have failed to get enough. In sections where much buckwheat is raised, late swarms do more towards filling their hives than where there is none.

MORE CARE NEEDED BY AFTER SWARMS WHEN HIVED.

Should it be thought best to hive after swarms, and risk the chances, they should receive a little extra attention after the first week or two, to destroy the worms; a little timely care may prevent considerable injury. They are apt to construct more combs in proportion to the number of bees, than others; consequently, such combs cannot be properly covered and protected. The moth has an opportunity to deposit her eggs on them, and, sometimes, entirely destroy them.

TWO MAY BE UNITED.

Whenever these swarms issue near enough together, it is best to unite them. I have said second swarms were generally half as large as the first. By this rule, two second swarms would contain as many bees as a first one, and four of the third, or one of the second issue, and two of the third, &c. If the first and second are of the ordinary size, I think it advisable always to return the third. But in large apiaries it is common for them to issue without any previous warning, just when a first one is leaving, and crowd themselves into their company, and seeming to be as much at home as though they were equally respectable.

Whenever the hives containing our swarms are full or very near it, the boxes should be put on without delay, unless the season of honey is so nearly gone as to make it unnecessary.

I have found it an advantage to hive a few of these very small swarms, on purpose to preserve queens, to supply some old stocks that sometimes lose their own at the extreme end of the swarming season. The cases to be mentioned at the last of the next chapter. I try and save one for about every twenty stocks that have swarmed.

CHAPTER XIV.

LOSS OF QUEENS.

OF SWARMS THAT LOSE THEIR QUEEN.

Swarms that lose their queen the first few hours after being hived, generally return to the parent stock; with the exception that they sometimes unite with some other. If much time has elapsed before the loss, they remain, unless standing on the same bench with another. On a separate stand they continue their labor, but a large swarm diminishes rapidly, and seldom fills an ordinary-sized hive. One singular circumstance attends a swarm that is constructing combs without a queen. I have never seen it noticed by any one, and may not always be the case, but *every* instance that has come under my notice, I have so found it. That is, four-fifths of the combs are drone-cells; why they thus construct them is another subject for speculation, from which I will endeavor in this instance to refrain.

A SUGGESTION AND AN ANSWER.

It has been suggested as a profitable speculation, "to hive a large swarm without a queen, and give them a piece of brood-comb containing eggs, to rear one, and then as soon as it is matured, deprive them of it, giving them another piece of comb, and continue it throughout the summer, putting on boxes for surplus honey. The bees having no young brood to consume any honey, no time will be lost, or taken to nurse them, and as a consequence they will be enabled to store large quantities of surplus honey."

This appears very plausible, and to a person without experience somewhat conclusive. If success depended on some animal whose lease of life was a little longer, it would answer better to calculate in this way. But as a bee seldom sees the anniversary of its birthday, and most of them perish the first few months of their existence, it is bad economy. It will be found that the largest amount of our surplus honey is obtained from our prolific stocks. Therefore it is all-important that every swarm and stock has a queen to repair this constant loss.

A DISPUTED QUESTION.

We now approach another disputed point in natural history, relative to the queen leaving at any time except when leading out a swarm. Most writers say that the young queen leaves the hive, and meets her paramour, the drone, on the wing. Others deny this *positively*, having watched a whole summer without seeing her highness leave. Consequently they have arrived at the very plausible and apparently consistent conclusion, that nature never intended it to be so, since it must happen at a time when the existence of the whole family depends entirely on the life of the queen. The stock at such times contains no eggs or larvæ, from which to rear another, if she should be lost. "The chances at such times of being devoured by birds, blown away by the winds, and other casualties, are too many, and it is not probable the Creator would have so arranged it." But facts are stubborn things; they will not yield one jot to favor the most "finely-spun hypothesis;" they are most provokingly obstinate, many times. When man, without the necessary observation, takes a survey through animated nature, and finds with scarcely an exception that male and female are about equal in number, he is ready, and often does conclude that one bee among thousands cannot be the only one capable of reproduction or depositing eggs. Why, the idea is preposterous! And yet only a little observation will upset this very consistent and analogous reasoning. So it appears to be with the excursions of the young queens. I was compelled, though reluctantly, to admit that they leave the hive. That their purpose is to meet the drones, I cannot at present contradict. Also, that, when the queen is once impregnated, it is operative for life, (yet it is another anomaly), as I never detected her coming out again for that purpose. What then is the use of the ten thousand drones that never fulfil this important duty? It seems, indeed, like a useless waste of labor and honey, for each stock to rear some twelve or fifteen hundred, when perhaps but one, sometimes not any of the whole number is of any use. If the risk is great in the queen's leaving, we find it arranged admirably in its not being too frequent.

A MULTITUDE OF DRONES NEEDED.

Instinct teaches the bee to make the matters left to them as nearly *sure* as possible. When they want one queen, they raise half a dozen. If one drone or only half a dozen were reared, the chances of the queen meeting one in the air would be very much reduced. But when a thousand are in the air instead of one, the chances are a thousand times multiplied. If a stock casts a swarm, there is a young queen to be impregnated, and be got safely back, or the stock is lost. Every time she leaves, there is a chance of her being lost, (one in fifteen). If the number of drones was any less than it is, the queen would have to repeat her excursions in proportion, before successful. As it is, some have to leave several times. The chances and consequences are so great, that on the whole no doubt but it is better to rear a thousand unnecessarily, than to lack one just in time of need. Therefore let us endeavor to be content with the present arrangement, inasmuch as we could not better it, and probably had we been consulted, would have so fixed "the thing, that it would not go at all."

But what is the use of the drones in hives that do not swarm, and do not intend it, situated in a large room or very large hives? Such circumstances seldom produce swarms, yet as regular as the return of summer, a brood of drones appear. What are they for? Suppose the old queen in such hive dies, leaving eggs or young larvæ, and a young queen is reared to supply her place. How is she to be impregnated without the drones? Perhaps they are taught that whenever they can afford it, they should have some on hand to be ready for an emergency. I have already said when bees are numerous, and honey abundant, they never fail to provide them. I once put a swarm in a glass hive. The queen was a cripple, having lost one of her posterior legs; in two months after she was replaced by one young and perfect. Here was an instance of drones being needed, when no intention of swarming was indicated; the hive was but little more than half full.

THE QUEEN LIABLE TO BE LOST IN HER EXCURSIONS.

This excursion of the queen, whenever I have witnessed it, always took place a little after the middle of the day, when the drones were out in the greatest numbers. At such times I have seen them leave amid rather more commotion than usual among the workers. I have watched their return, which varied from three minutes to half an hour, and seen them hover around their own hive, apparently in doubt whether they belonged in that, or the next; in a few instances they have actually settled on the neighboring hive, and would have there perished, but for my assistance in putting them right.

THE TIME WHEN IT OCCURS.

Thus we see that queens are lost on these occasions from some cause, and part of them by entering the wrong hive, perhaps most of them; if so, it is another good reason for not packing stocks too close. The hives are very often nearly alike in color and appearance. The queen coming out for the first time in her life, is no doubt confused by this similarity.

The number of such losses in a season has varied: one year the average was one in nine, another it was one in thirteen, and another one in twenty. The time from the first swarm also varies from twelve to twenty days. The inexperienced reader should not forget that it is the old stocks which have cast swarms, where these accidents happen; the old queen having left with the first swarm. Also all after swarms are liable to the same loss. I would suggest that these have abundant room given between the hives; if it is necessary to pack close, let it be the first swarms, where the old queen has no occasion to leave. Having never seen this matter fully discussed, I wish to be somewhat particular, and flatter myself that I shall be able to direct the careful apiarian how to save a few stocks and swarms annually, that is, if he keeps many. A few years ago, I wrote an article for the Albany Cultivator. A subscriber of that paper told me a year afterwards that he saved two stocks the next summer by the information; they were worth at least five dollars each, enough to pay for his paper ten years or more.

When a stock casts but one swarm, the queen having no competitors to interfere with her movements, will leave in about fourteen days, if the weather is fair; but should an after swarm leave, the oldest of the young queens will probably go with that, of course: then, it must be later before the next is ready: it may be twenty days, or even more; those with after swarms will vary from one to six. It *always must* occur when no eggs or larvæ exist, and no means left to repair this loss; a loss it is, and a serious one; the bees are in as much trouble as their owner, and a great deal more, they seeming to understand the consequences, and he, if he knows nothing of the matter, has no trouble. Should he now, for the first time, learn the nature of it, he will at the same time understand the remedy.

INDICATIONS OF THE LOSS.

The next morning after a loss of this kind has occurred, and occasionally at evening, the bees may be seen running about in the greatest consternation, outside, to and fro on the sides. Some will fly off a short distance and return; one will run to another, and then to another, still in hopes, no doubt, of finding their lost sovereign! A neighboring hive close by, on the same bench, will probably receive a portion, which will seldom resist an accession under such circumstances. All this will be going on while other hives are quiet. Towards the middle of the day, this confusion will be less marked; but the next morning it will be exhibited again, though not so plainly, and cease after the third, when they become apparently reconciled to their fate.

They will continue their labors as usual, bringing in pollen and honey. Here I am obliged to differ with writers who tell us that all labor will now cease. I hope the reader will not be deceived by supposing that because the bees are bringing in pollen, that they *must* have a queen; I can assure you it is not always the case.

THE RESULT.

The number of bees will gradually decrease, and be all gone by the early part of winter, leaving a good supply of honey, and an extra quantity of bee-bread, as before mentioned, because there has been no young brood to consume it. This is the case when a large family was left at the time of the loss. When but few bees are left, it is very different; the combs are unprotected by a covering of bees; the moth deposits her eggs on them, and the worms soon finish up the whole. Yet the bees from the other stocks will generally first remove the honey.

AGE OF BEES INDICATED.

Hundreds of bee-keepers lose some of their stocks in this way, and can assign no reasonable cause. "Why," say they, "there wasn't twenty bees in the hive; it was all full of honey," or worms, as the case may be. "Only a short time before, it was full of bees; I got three good swarms from it, and it always had been first rate, but all at once the bees were gone. I don't understand it!" Such bee-keepers cannot understand how rapidly a family of bees diminish, when there is no queen to replenish with young this mortality of the old ones. I doubt whether the largest and best family possibly could be made to exist six months, without a queen for their renewal, except, perhaps, through the winter.

When standing close on one bench, they are gone sooner than if on separate stands, as they often join a neighboring hive when they can walk to it.

NECESSITY OF CARE.

As this tumult cannot be seen but a few days at most, it is well—yes, it is necessary—to make it a duty to glance at the hives at this period after swarming, *every morning*; a glance is sufficient to tell you of the fact. Remember to reckon from the date of the first issue; this occurs when the first royal cells are sealed over, and is the best criterion as to when the queen will leave. If the first swarm issue and return, it can make no difference; reckon from their first issuing.

REMEDY.

When you discover a loss, first ascertain if there is any after swarm to be expected from another stock, (by listening for the piping); if so, wait till it issues, and obtain a queen from that for your stock; even if there is but one, take it, and let the bees return; they would be likely to come out again the next day; if not, it is very often no great loss.

Should no such swarm be indicated, go to a stock that has cast a first swarm within a week; smoke it and turn it over, as before directed, find a royal cell, and with a broad knife cut it out, being careful not to injure it. This must now be secured in the other hive in its natural position, the lower end free from any obstacle, that would interfere with the queen leaving it. It will make but little difference whether at the top or bottom, providing it is secure from falling.

I generally introduce it through a hole in the top, taking care to find one that will allow the cell to pass down between two combs. It being largest at the upper end, the combs each side will sustain it, and leave the lower end free. In a few hours the bees will secure it permanently to the combs with wax. This operation cannot be performed in a chamber hive, as it is impossible to see the arrangement of the combs through the holes. To put it in at the bottom is some more trouble; the difficulty is, to fasten it, and prevent it resting on the end. I have done it as follows: Get an *old* thick piece of dry comb some three inches square; cut out an inch of the middle. At right angles with this, in one edge in the centre, make another to intersect it, just the size of the cell, and have the lower end reach into the opening. This comb will keep it in the right position, and may rest on the floor-board. It can now be put in the hive, cutting out a piece of comb to make room for it if necessary.

Soon after such cell is introduced, the bees are quiet. In a few days it hatches, and they have a queen as perfect as if it had been one of their own rearing. This queen of course will be necessitated to leave the hive, and will be just as liable to be lost, but no more so than others, and must be watched the same. It is unnecessary to look for a cell in a stock that has cast its first swarm more than a week before, as they are generally destroyed by that time, (sometimes short of it,) unless they intend to send out an after swarm.

MARK THE DATE OF SWARMS ON THE HIVE.

Should you have so many stocks that you cannot remember the date of each swarm without difficulty, it is a good plan to mark the date on one side or corner of the hive, as it issues. You can then tell at once where to look for a cell when wanted.

It will sometimes happen that a queen may be lost at the extreme end of the swarming season, when no other stock contains such cells. I then look around for the poorest stock or swarm that I have on hand, one that I can afford to sacrifice, if it possesses a queen, to save the one that has sustained this loss; this is not often the case, but is sometimes. I have a few times put just bees enough with the queen to keep her in a box, and kept them for this purpose, as was mentioned in the last chapter. When introduced, the bees are generally killed, but the queen is preserved.

OBTAINING A QUEEN FROM WORKER BROOD.

There is yet another method to be adopted, and that is, to obtain a piece of brood-comb containing workers' eggs, or larvæ very young. You will generally find it without much trouble, in a young swarm that is making combs; the lower ends usually contain eggs; take a piece from one of the middle sheets, two or three inches long, (you will probably use smoke by this time without telling). Invert the hive that is to receive it, put the piece edgewise between the combs, if you can spread them apart enough for the purpose; they will hold it there, and then there will be ample room to make the cells. They will nearly always rear several queens. I have counted nine several times, which were all they had room for. But yet I have very little confidence in such queens, they are almost certain to be lost.

THEY ARE POOR DEPENDENCE.

Therefore I would recommend getting a royal cell whenever it is practical. There is yet another advantage; you will have a queen ready to lay eggs two or three weeks earlier, than when they are compelled to commence with the egg. I have put such piece of brood-comb in a small glass box on the top of the hive instead of the bottom, because it was less trouble, but in this case the eggs were all removed in a short time; whether a queen was reared in the hive or not I cannot say; but this I know, I never obtained a prolific queen, after repeated experiments in this way.

It would appear that I have been more unfortunate with queens reared in this way, than most experimenters. I have no difficulty to get them formed to all appearance perfect, but lose them afterwards. Now whether this arose from some lack of physical development, by taking grubs too far advanced to make a perfect change, or whether they were reared so late in the season, that most of the drones were destroyed, and the queen to meet one had to repeat her excursions till lost, I am yet unable to *fully* determine. To test the first of these questions, I have a few times removed all the larvæ from the comb; leaving nothing but eggs, that all the food given them might be "royal pap," from the commencement, and had no better success so far. Yet occasionally prolific queens have been reared when I could account for their origin in no other way but from worker eggs. But you will find they are not to be depended upon generally.

Sometimes, after all our endeavors, a stock or two will remain destitute of a queen. These, if they escape the worms, will generally store honey enough in this section to winter a good family. This will have to be introduced, of course, from another hive, containing a queen; but this belongs to Fall management.

As respects the time that elapses from the impregnation of the queen till the commencement of egg laying, I cannot tell, but guess it might be about two or three days. I have driven out the bees twenty-one days after the first swarm, when no second swarm had issued—the young queen came out on the fourteenth day. I found eggs and some very young larvæ. When it is remembered that eggs remain three days before they hatch, it shows that the first of these must have been deposited some four or five days. When writers tell us the exact time to an hour (46 or 48) from impregnation to laying, I am willing to admit the thing in this case, but feel just as if I would like to ask how they managed to find out the fact; by what sign they knew when a queen returned from an excursion, whether she had been successful or not, in her amours; or, whether another effort would have to be made; and then, how they managed to know exactly when the first egg was laid.

Occasionally a queen is lost at other than the swarming season, averaging about one in forty. It is most frequent in spring; at least it is generally discovered then. The queen may die in the winter, and the bees not give us any indications till they come out in spring. (Occasionally they may all desert the hive, and join another.) If we expect to ascertain when a queen is lost at this season, we must notice them just before dark on the first warm days—because the mornings are apt to be too cool for any bees to be outside—any unusual stir, or commotion, similar to what has been described, shows the loss. This is the worst time in the year to provide the remedy, unless there should happen to be some very poor stock containing a queen, that we might lose any way—then it might be advisable to sacrifice it to save the other, especially if the last contained all the requisites of a good stock except a queen. Some eight or ten, that I have managed in this way, have given me full satisfaction. I have at other times let them go till the swarming season, and then procured a queen, or introduced a small swarm; at which time they are so reduced as to be worth but little, even when not affected by the worms. To obviate this loss in this way, it might be an advantage to transfer the bees to the next stock, if it was not too full already; or the bees of the next stock to this. Let the age and condition of the combs, quantity of stores, &c., decide.

CHAPTER XV.

ARTIFICIAL SWARMS.
PRINCIPLES SHOULD BE UNDERSTOOD.

Artificial swarms can be made with safety at the proper season. To the bee-keeper who wishes to increase his stocks, it will be an advantage to understand some of the principles. I have had some little experience that has led to different conclusions from those of some others. I have seen it stated, and found the assertion repeated by nearly every writer, that "whenever bees were deprived of their queen, if they only possessed eggs or young larvæ, they would not fail to rear another," &c. There are numerous instances of their doing this, but it is not to be depended upon, especially when left in a hive full of combs, as the following experiments tend to prove.

SOME EXPERIMENTS.

Several years since I had a few stocks well supplied with bees, and every indication of swarming present, such as clustering out, &c., but they pertinaciously adhered to the old stock, through the whole swarming season! Others apparently not as well supplied with bees threw off swarms. I had but few stocks, and was very anxious to increase the number; but these were provokingly indifferent to my wishes. Taking the assertions of these authors for facts, I reasoned thus: In all probability there are eggs enough in each of those stocks. Why not drive out a portion of the bees, with the old queen, and leave about as many as if a swarm had issued? Those left will then raise a queen, and continue the old stock, and I shall have six instead of the three, that have been so obstinate. Accordingly, I divided each, examined and found eggs and larvæ. Of course all *must be right*. Now, thought I, my stocks can be doubled at least annually. If they do not swarm, I can drive them.

THE RESULT UNSATISFACTORY.

My swarms prospered, the old stocks seemed industrious, bringing in pollen in abundance, which to me at *that* time, was conclusive that they had a queen, or soon would have. I continued to watch them with much interest, but somehow, after a few weeks, there did not seem to be quite as many bees; a few days later, I was quite *sure* there was not. I examined the combs, and behold there was not a cell containing a young bee of any age, not even an egg in any one of these old stocks. My visionary anticipations of future success speedily retrograded about this time.

I had, it is true, my new swarms in condition to winter, although not quite full; but the old ones were not, and nothing was gained. I had some honey, a great deal of bee-bread and old black comb. Had I let them alone, and put on boxes, I should have probably obtained twenty-five or thirty pounds of pure honey from each, worth five times as much as what I did get; besides, the old stocks, even with the old comb, would have been better supplied with both honey and bees; altogether much better, as stocks for wintering. Here was a considerable loss, merely by not understanding the matter.

I carefully looked the bees over, and ascertained to a certainty that neither of them had a queen. I smothered what few there was left in the fall. I then knew of no better way. I had been told that the barbarous use of fire and brimstone was part of the "luck;" that a more benevolent system would cause the bees "to run out," &c.

FURTHER EXPERIMENTS.

Subsequent to these experiments, I thought perhaps the jarring of the hives in driving might have some effect on the bees, and prevent their rearing a queen. This idea suggested the dividing hive, when the division could be made quietly; but success was yet uncertain. I was told to confine the bees in the old stock twenty-four hours or more, after driving out a swarm; this I tried, with no better results. Again, I drove out the swarm, looked out the queen, and returned her to the old stock, compelling the new swarm to raise one. To be certain they did so, I constructed a small box about four inches square, by two in thickness; the sides glass. In this I put the piece of brood-comb containing eggs and larvæ, and then put it on the hive containing the swarm, having holes for communication, a cover to keep it dark, &c. They were very sure to rear queens, but from some cause were lost after they were matured.

Now, if others have been more successful in these experiments than myself, it indicates that some favorable circumstances attended them that did not me. I have not the least doubt but the result will be favorable sometimes. Yet from the foregoing, I became satisfied that not one of these methods could be relied upon. Instead of constructing a queen's cell, and then removing the egg or larva to it from another cell, I always found that the cell containing such egg or larva was changed from the horizontal to the perpendicular; such cells as were in the way below were cut off, probably using the material in forming one for royalty, which, when finished, contains as much material as fifty or a hundred others.

My experiments did not end here. I can now make artificial swarms, and succeed nine times in ten with the first effort, and the reader can as easily do the same. It must be in the swarming season, or as soon as the first regular swarm issues. You want some finished royal cells that any stock having cast a swarm will furnish, (unless in rare instances, where they are too far up among the combs to be seen.)

A SUCCESSFUL METHOD.

When you are all ready, take a stock that can spare a swarm; if bees are on the outside, raise the hive on wedges, and drive them in with a little water, and disturb them gently with a stick. Now smoke and invert it, setting the empty hive over. If the two hives are of one size, and have been made by a workman, there will be no chance for the bees to escape, except the holes in the side; these you will stop; (no matter about a sheet tied around it.) With a light hammer or stick, strike the hive a few times lightly, and then let it remain five minutes. This is very essential, because most of the bees, if allowed the opportunity, will fill themselves with honey after such disturbance.

All regular swarms go forth so laden. A supply is necessary when bad weather follows soon after. It is also used in forming wax, a very necessary article in a new hive. The amount of honey carried out of a stock by a good swarm, together with the weight of the bees (which is not much), will vary from five to eight pounds.

This, allowing time for the bees to fill their sacks, and supplying the old stock with a royal cell, I believe is entirely original: the importance of which the reader can judge.

ADVANTAGES OF THIS METHOD.

It is very plain that a queen from such finished cell must be ready to deposit eggs several days sooner than by any other method that we can adopt. It is also clear that if we have a dozen queens depositing eggs by the 10th of June, that our bees are increasing faster, on the whole, than if but half that number are engaged in it for a month later. There is yet another advantage. The sooner a young queen can take the place of the old one in maternal duties, the less time will be lost in breeding, the more bees there will be to defend the combs from the moth, and the surest guaranty for surplus honey.

When the bees have filled their sacks, proceed to drive them into the upper hive by striking the lower one rapidly from five to ten minutes. A loud humming will mark their first movement. When you think half or two-thirds are out, raise the hive and inspect progress. They are not at all disposed to sting in this stage of proceeding, even when they escape outside. If full of honey, they are seldom provoked to resentment. The only care will be not to crush too many that get between the edges of the hives. The loud buzzing is no sign of anger. If your swarm is not large enough, continue to drive till it is. When done, the new hive should be set on the stand of the old one. A few minutes will decide whether you have the queen with the swarm, as they remain quiet: otherwise uneasy, and run about, when it will be necessary to drive again.

If both hives are one color, set the old one two feet in front; but if of different colors, a little more. I prefer this position to setting the old stock on one side, even when there is room; yet it can make but little difference. Should you set it on one side, let the distance be less. When the old stock is taken much farther than this rule, all the bees that have marked the location (and all the old ones will have done so) will go back to the old stand, and none but young bees that have never left home will remain. The same will be the case with the new swarm if moved off. It will not do to depend on the old queen keeping them, as she does when they swarm out naturally. This has been my experience. Try it, reader, and be satisfied, by putting either of the hives fifteen or twenty feet distant.

Before you turn over the old stock, look among the combs as far as possible for queens' cells; if any contain eggs or larvæ, you may safely risk their rearing a queen; but otherwise wait till next morning, or at least twenty-four hours, then go to a stock that has cast a swarm, and obtain a finished royal cell, as before directed, and introduce it. You will have a queen here as soon as if it had been left in the original hive, and no risk of an after swarm, because there is but one. But when there are young queens in the cells at the time of driving, after swarms may issue. Should a queen-cell be introduced immediately, it is more liable to be destroyed than after waiting twenty-four hours; and then is not always safe. After it has had time to hatch, (which is about eight days after being sealed), cut it out, and examine it: if the lower end is open, it indicates that a perfect queen has left it, and all is safe; but if it is mutilated or open at the side, it is probable that the queen was destroyed before maturity, in which case, another cell will have to be given them.

ARTIFICIAL SWARMS ONLY SAFE NEAR THE SWARMING SEASON.

By what I have said about artificial swarms, it would appear that it is unsafe at any time but the swarming season; that is my opinion. It may do a little in advance or a little after, providing royal cells can be had. By feeding as directed, (in Chapter IX.) you may induce a stock to send out a swarm some days in advance of the regular season, thereby giving you a chance for these cells somewhat early.

SOMETIMES HAZARDOUS.

To make such swarms at any time when the bees are destroying drones, would be extremely hazardous, not only on account of the young queen being impregnated, but their massacre denotes a scarcity of honey. Therefore I would advise never to make swarms, or drive out bees at such periods, when it can be avoided, without spare honey is on hand to feed them.

SOME OBJECTIONS.

It has been argued by some, and with much reason, that "nature is the best guide, and it is better to let the bees have their own way about swarming—if honey is abundant, and the stock is in condition to spare a swarm, their own instincts will teach them to construct royal cells; if it fails before they are ready, and the royal brood is destroyed, it is because the existence of the swarm would be precarious, and it is best not to issue." I will grant that in many instances it is better. The chance is better for surplus honey; the stock is quite sure to be in condition to winter; and some judgment is required to tell when a stock can spare a swarm.

But yet, we are sometimes anxious to increase our stocks to the utmost that safety will allow, and often have some that can spare a swarm as well as not, but refuse to leave; perhaps commence preparations, and in a few days abandon them. Now it is evident that as long as many continue such preparation, that honey is sufficiently abundant to put the safety of the swarm beyond hazard; some stocks will swarm while these others just as good, (that had abandoned it before) and have not now begun again, to be in time before a partial failure of honey, and some may not have commenced in season.

NATURAL AND ARTIFICIAL SWARMS EQUALLY PROSPEROUS.

I can see no difference in artificial or natural swarms of equal size, at the same time. By taking the matter in time into our own hands, with the rules given, we make a sure thing of it, that is, we are sure to get the swarms, when if left to the bees it would be uncertain, and no greater risk afterwards than with natural issues.

THIS MATTER TOO OFTEN DELAYED.

I am aware that this matter will be apt to be put off too long; "wait and see if they don't swarm," will be the motto of too many, and when the season is over, drive them. Perhaps a good swarm has set outside the hive, all through the best of the honey season, and done nothing, while they could have half filled a hive; but this is all lost now, as well as the best chances for getting cells. Let me impress the necessity of doing it in season, when it will pay. If you intend to have a swarm from every stock that can spare one, begin when nature points out the proper time, which is, when the regular ones begin to issue. It must, indeed, be a poor season when there are none.

IS THE AGE OF THE QUEEN IMPORTANT?

There is another object effected in this way, considered by some apiarians as very important. It is the change of the queens in the old stock. A young queen is thought to be "much more prolific than an old one." They even recommend keeping none "over two or three years old," and give directions how they may be renewed. But as I have been unable to discover any difference in relation to the age in this respect, I shall not at present take much time to discuss it. It is well enough, when we can take our choice without trouble, to preserve a young queen. When we consider that there are but few queens but what will deposit three times as many eggs in a season as are matured, it looks as if it would hardly pay to take much trouble to change them. At what time the queen becomes barren from old age, I presume has never yet been fully determined.

A friend of mine has had a stock in a large room eight years, that has never swarmed, and is still prosperous! I think it very probable that this queen will gradually decay, and possibly become barren, some weeks before she dies; if so, this stock will soon die off. A few such cases will probably occur in swarming hives, perhaps one in fifty, but generally such old and feeble queens are lost when they leave with the swarm, especially in windy weather. As long as they are able to go with the swarm, and sometimes when they are not, I have found them sufficiently prolific for all purposes. I would rather risk their fecundity, and hive the swarm, than to allow the bees to return to the parent stock, and wait eight or nine days for a young queen to mature. A great many will remain idle, even if there is room to work in the boxes.

CHAPTER XVI.

PRUNING.

Notwithstanding I have given the method of pruning in the chapter on hives, (page 23, Chapter II.) it will be necessary to give the tyro in bee-culture a few more particulars. The season for doing it is of importance.

DIFFERENT OPINIONS AS TO TIME.

The month of March has been recommended by several; others prefer April, August, or September. Here, as usual, I shall have to differ from them all, preferring still another period, for which I offer my reasons, supposing, of course, that the reader is conscious of a freeman's privilege, that is, to adopt whatever method he thinks proper, on this, as on any other point.

ANOTHER TIME PREFERRED.

There is but one period from February till October, when prosperous stocks are free from young brood in the combs. If combs are taken out when occupied, there must be a loss of all the young bees they contain; which may be avoided. The old queen leaves with the first swarm; all the eggs she leaves in the worker-cells will be matured in about twenty-one days, consequently this is the time to clear out the old combs with the least waste. A few drones will be found in the cells, that would require a few days more to hatch, but these are of no account. Also a few very young larvæ and some eggs may be sometimes found, the product of the young queen; these few must be wasted, but as the bees have expended no labor upon them as yet, it is better to sacrifice these than the greater number left by her mother, which have consumed their portion of food; the bees have sealed them up, and now only require the necessary time to mature, to make a valuable addition to the stock.

SHOULD NOT BE DELAYED.

Should this operation be put off for a time much longer than three weeks, the young queen will so fill the combs again as to make it a serious loss. Therefore, I wish to urge strongly attention to this point at the proper season. If you think it unimportant to mark the date of your first swarms for the purposes mentioned in another place, it will be found very convenient here, for those that need pruning.

It is also recommended by some, to take only a part, say one-third or half, in a season; thereby taking two or three years to renew the combs. This is advisable only when the family is very small. As this space made by pruning cannot be filled without wax and labor, our surplus honey will be proportionate to its extent. Now suppose we take out half the old combs, and get half a yield of box honey this year, and the same next, or make a full operation of it and get none this year, and a full one next. What is the difference? There is none in point of honey, but some in trouble, and that is in favor of a full operation at once. We have to go through with about the same trouble to get one-third or half as to take the whole.

OBJECTION TO PRUNING.

The objection to this mode of renewing combs generally, will be the fear of getting stung. But I can assure you there is but little danger, not as much as to walk among the hives in a warm day. Only begin right, use the smoke, and work carefully, without pinching them, and you will escape unhurt generally.

STOCKS PRUNED NOW ARE BETTER FOR WINTER.

Besides the advantage of saving a large brood by pruning at this season, such stocks will usually refill before fall, and are much better for wintering, which is not the case when it is done later. We must of necessity then waste the brood, and have a large space unoccupied with combs through the winter. But few combs can then be made, and those few must be at the expense of their winter stores, unless we resort to feeding.

These objections apply with greater force to pruning in March or April. The loss of brood is of much more consequence now, than in mid-summer, or even later, and a space to be filled with combs is a serious disadvantage. It is important that the bees should devote their whole attention now to rearing brood, and be ready to cast their swarms as early as possible. One *early* swarm is worth two late ones. Suppose a stock, instead of collecting food and nursing its young, is compelled to expend its honey and labor in secreting wax and constructing combs before it can proceed with breeding advantageously, it *must of necessity* be some weeks later.

Further, I have always found it best to have the bees out of the way, during this operation. It will be found much more difficult to drive the bees out of a hive in the cool weather of March or April, than in summer, as they seem unwilling to shift their warm quarters and go into a cold hive.

It is presumed the reader will bear in mind the disadvantages already given of too frequently renewing combs; the little value of combs for storing honey, *for our use*, after being once used for breeding; the necessity of the bees using them as long as they possibly will answer; and not compel them to be filling the hive, when they might be storing honey of the purest quality in boxes, &c.

Vide remarks on this subject on page 22, Chapter II.

CHAPTER XVII.

DISEASED BROOD.

This, like many other chapters in this work, is probably new, as I, never saw one thus headed. A few newspaper discussions are about all that have yet appeared on this subject.

NOT GENERALLY UNDERSTOOD.

This disease is probably of recent origin. Mr. Miner, it appears, knew nothing of it until he moved from Long Island to Oneida County, in this State. Mr. Weeks, in a communication to the N.E. Farmer, says, "Since the potato rot commenced, I have lost one-fourth of my stocks annually, by this disease;" at the same time adds his fears, that "this race of insects will become extinct from this cause, if not arrested." (Perhaps I ought to mention, that he speaks of it as attacking the "chrysalis" instead of the larva; but as every thing else about it agrees exactly, there is but little, doubt of its being all one thing.)

MY OWN EXPERIENCE.

My first experience will probably go back to a date beyond many others; it is almost twenty years since the first case was noticed. I had kept bees but four or five years when I discovered it in one of my best stocks; in fact, it was No. 1 in May and first of June. It cast no swarm through the summer; and now, instead of being crowded with bees, it contained but very few; so few, that I dared not attempt to winter it. What was the matter? I had then never dreamed of ascertaining the condition of a stock while there were bees in the way, but was like the unskilful physician who is obliged to wait for the death of his patient, that he may dissect and discover the cause. I accordingly consigned what few bees there were to the "brimstone pit."

DESCRIPTION OF DISEASE.

A "*post mortem*" examination revealed the following circumstances: Nine-tenths of the breeding-cells were found to contain young bees in the larva state, stretched out at full length, sealed over, dead, black, putrid, and emitting a disagreeable stench. Now here was one link in the chain of cause and effect. I learned why there was a scarcity of bees in the hive. What should have constituted their increase, had died in the cells; none of them were removed, consequently but few cells, where any bees could be matured, were left.

THE CAUSE UNCERTAIN.

But when I attempted the next link in the chain (to wit) What caused the death of this brood just at this stage of development? I was obliged to stop. Not the least satisfaction could be obtained. All inquiries among the bee-keepers of my acquaintance were met with profound ignorance. They had "never heard of it!" No work on bees that I consulted ever mentioned it.

Subsequently, I had more stocks in the same situation. I found, whenever the disease existed to any extent, that the few bees matured were insufficient to replace those that were lost; that the colony rapidly declined, and *never afterwards cast a swarm*!

REMEDIAL EXPERIMENTS.

As for remedies, I tried pruning out all those combs containing brood, leaving only such as contained honey, and let the bees construct new for breeding. It was "no use," these new combs were invariably filled with diseased brood! The only thing effectual was to drive out the bees, into an empty hive. In this way, when done in season, I generally succeeded in rearing a healthy stock. But here was a loss of all surplus honey, and a swarm or two that might have been obtained from a healthy one.

PUBLIC INQUIRY AND ANSWERS.

I had so many cases of the kind, that I became somewhat alarmed, and made inquiry through the Cultivator, (an agricultural paper,) as to a cause, and remedy, offering a "reward for one that would not fail when thoroughly tested," &c. Mr. Weeks, in answer, said, "that cold weather in spring chilling the brood was the cause." (This was several years prior to his article in the N.E. Farmer.) Another gentleman said, "dead bees and filth that accumulated during winter, when suffered to remain in the spring, was the cause." A few years after, another correspondent appeared in the Cultivator, giving particulars of his experience, proving very conclusively to himself and many others, that cold was the cause. Having mislaid the paper containing his article, I will endeavor to quote correctly from memory. He had "three swarms issue in one day; the weather during the day changed from very hot to the other extreme, producing frost in many places the next morning. These swarms had left but few bees in the old stocks, and the cold forced them up among the combs for mutual warmth; the brood near the bottom, thus left without bees to protect it with animal heat, became chilled, and the consequence was diseased larvæ." He then reasoned thus: "If the eggs of a fowl, at any time near the end of incubation, become chilled from any cause, it stops all further development. Bees are developed by continued heat, on the same principle, and a chill produces the same effect, &c.; afterwards, other swarms issued under precisely similar circumstances; but these old stocks were covered with a blanket through the night, which enabled the bees to keep at the bottom of the hive. In a few days, enough were hatched to render this trouble unnecessary. These last remained healthy." He further says, that "last spring was the first time I ever knew them to become diseased before swarming had thinned the population. The weather was remarkably pleasant through April. The bees obtained great quantities of pollen and honey, and by this means extended their brood further than usual at this season. Subsequent chilly weather in May, caused the bees to desert a portion of brood, which were destroyed by the chill."

Now this is reasoning from cause to effect very consistently.

ANSWERS NOT SATISFACTORY.

Had I no experience further than this, I should, perhaps, rest satisfied as to the cause, and should endeavor to apply the remedy. Several other writers have appeared in different papers, on this subject, and nearly all who assign a cause have given this one as the most probable. Now I have known the chrysalis in a few stocks to be chilled and destroyed by a sudden turn of cold weather, yet these were removed by the bees soon after, and the stocks remained healthy. To me the cause assigned appears inadequate to produce *all* the results with the larvæ. After close, patient observation of fifteen years, I have never yet been wholly satisfied that any one instance among my bees, was thus produced.

A CAUSE SUGGESTED.

We are all familiar to some extent with the contagious diseases of the human family, such as small-pox, whooping-cough, and measles, and their rapid spread from a given point, &c. We must also admit that some cause or causes, adequate to the effect, must have produced the first case. To contagion, then, I would attribute the spread of this disease of our bees, at least nineteen cases in twenty. I will admit, if you please, that one stock in twenty or fifty may be somewhat affected by a chill to a small extent. It is only a portion of the brood that is in danger—only such as have been sealed over, and before they have progressed to the chrysalis state, are attacked. How many then can there be in a hive at any one time, in just the right stage of development to receive the fatal chill? Of course there will be some; but they should be confined to the cells near the bottom, where the bees had left them exposed. These should be all; and these few would never seriously damage the stock. Why then does this disease, when thoroughly started, spread so rapidly throughout all the combs in the hive? Will it be said that the chill is repeated every few days through the summer? Or will it be admitted that something else may continue it?

I think there must be other causes, besides the chill, even to start it, in most cases. As our practice will be in accordance with the view we take of this matter, and the result of our course will be somewhat important, I will give some of the reasons that have led to this conclusion.

REASONS FOR THE OPINION.

For instance, I had all the bees of a good swarm leave the hive in March; after flying a time, they united with another good stock, making double the usual number of bees at this season; enough to keep the brood sufficiently warm at any time; if other stocks with half or a quarter of the number could. By the middle of June, the bees were much reduced, and had not cast a swarm. It was examined, and the brood was found badly diseased. My best and most populous stocks, in spring, are just as liable, and I might add more so, than smaller or weaker families. I have had two large swarms unite, and were hived together, that were diseased the next autumn. These cases prove strongly, if not conclusively, that animal heat is not the only requisite. The fact that when I had pruned out all affected comb from a diseased stock, and left honey in the top and outside pieces, and the bees constructed new for breeding, and the brood in such were invariably affected, though only a few at first, and increasing as the combs were extended; led me to suppose that it was a contagious disease, and the virus was contained in the honey. Some of it had been left in these stocks, and very probably the bees had fed it to the brood. To test this principle still further, I drove all the bees from such diseased stocks, strained the honey, and fed it to several young healthy swarms soon after being hived. When examined a few weeks after, every one, without an exception, had caught the contagion.

Here then is a clue to the cause of this disease spreading, whether we have its origin or not. We will now see if we can trace it through, if there is any consistency in its transfer from one stock to another.

CAUSE OF ITS SPREADING.

Suppose one stock has caught the infection, but a small portion of the brood is dead. In the heat of the hive, it soon becomes putrid; other cells adjoining with larvæ of the right age are soon in the same condition. All the breeding combs in the hive become one putrid mass, with an exception, perhaps, of one in ten, twenty or a hundred, that may perfect a bee. Thus the increase of bees is not enough to replace the old ones that are continually dying off. It is plain, therefore, that this stock *must* soon dwindle down to a very small family. Now let a scarcity of honey occur in the fields, this poor stock cannot be properly guarded, and is easily plundered of its contents by the others. Honey is taken that is in close proximity to dead bodies, corrupting by thousands, creating a pestilential vapor, of which it has probably absorbed a portion. The seeds of destruction are by this means carried into healthy stocks. In a short time, these in turn fall victims to the scourge; and soon dwindle away, when some other strong stock is able to carry off *their* stores; and only stop, perhaps, at the last stock! The moth is ever ready with her burden of eggs, which she now without hindrance deposits directly on the combs. In a short time the worms finish up the whole business, and are judged guilty of the whole charge; merely because they are found carrying out effects that speedily follow such causes.

Let the reader who doubts this theory, simply strain out honey, vitiated in this way, and feed it to a few stocks or swarms, that are healthy; and if they escape, communicate the fact to the public. But should he become satisfied that such honey is poison to his bees, he will with me, and all others interested, wish to stop this growing evil.

NOT EASILY DETECTED AT FIRST.

It is very difficult to detect the first hundred or two that die in a stock. But when nine-tenths of the breeding cells hold putrid larvæ, there is but very little trouble in making out a correct diagnosis. The bees are few and inactive. When passing the hive our olfactories are saluted with a nauseous effluvia, arising from this corrupting mass. Now, if we wish, or expect to escape, the most severe penalty, our neglect must never allow this extent of progression before such a stock is removed. Therefore, we must watch symptoms—ascertain the presence of the disease *at the earliest moment possible*.

SYMPTOMS TO BE OBSERVED.

As no part of the breeding season is exempt, the stocks should be carefully observed during spring, and fore part of summer, relative to increase of bees. When one or more is much behind others in this respect, make an examination immediately. (I would here urge again the convenience of the simple, common hive, over those more complicated, or suspended, and difficult to turn over. In one case we might make an examination in season; in the other, too much trouble and difficulty might cause it to be put off too long.) The hive must be inverted, and the bees smoked out of the way. Our attention is to be directed to the breeding cells; with a sharp-pointed knife, proceed to cut off the ends of some of them that appear to be the oldest; bearing in mind that young bees are always white, until some time after they take the chrysalis state. Therefore, if a larva is found of a dark color, it is dead! Should a dozen such be found, the stock should be condemned at once, and all the bees driven into an empty hive. (The directions for this have been given, see page 31.) If honey should be scarce, at the time, they should be fed.

SCALDING THE HONEY TO DESTROY THE POISON FOR FEEDING.

The honey from the old hive may be used, if you will only first destroy the virus. This, I have ascertained, may be done by scalding: add a half-pint of water to about ten lbs.; stir it well, and heat it to the boiling point, and carefully remove all the scum.

Stocks in which the disease has not progressed too far, will generally swarm.

WHEN TO EXAMINE STOCKS THAT HAVE SWARMED.

Three weeks from the first swarm, will be the time to examine them. I make it a rule to inspect all my stocks at this period. It is easily done now, as about all the healthy brood (except drones) should be matured in that time. By perseverance in these rules, I allow no stocks to dwindle away until they are plundered by others. If all my neighbors were equally careful, this disease would probably soon disappear. This is like one careless farmer allowing a noxious weed to mature seeds, to be wafted by winds on the lands of a careful neighbor, who must fortify his mind to continual vigilance, or endure the injury of a foul pest. So with the successful apiarian; in sections where the disease has appeared (it has not in all), he must be continually on the watch; it is the price of success.

CARE IN SELECTING STOCK HIVES FOR WINTER.

Again, after the breeding season is over, in the fall, *every stock should be thoroughly inspected, and all diseased ones condemned for stock hives*. It is better to do it, even if it should take the last one. It would pay much better to procure others instead, that are healthy.

Persons wishing to eat the honey from such hives, will experience no bad effects from it, if they are careful to remove all the dead brood, as they take it out of the hive.

The greatest distance that I ever knew bees to go, and plunder a defenceless stock of its contents, was three-fourths of a mile. Very likely they would go farther on some occasions, but not often.

ACCUSATIONS NOT ALWAYS RIGHT.

Careless bee-keepers, when their hives are thus robbed, feel regret, or are more often vexed at somebody—at the result of their carelessness. The person, keeping most bees in a neighborhood, must expect to be accountable for all effects of their ignorance, mismanagement, or carelessness, and consequent "bad luck;" when all the honey thus obtained, probably carries with it more mischief than can be eradicated in a twelvemonth, thereby giving the real cause of complaint to the other party.

CHAPTER XVIII.

IRRITABILITY OF BEES.

Keeping bees good-natured, offers a pretty fair subject for ridicule: it seems rather too absurd to teach *a bee* anything! Nevertheless, it is worth while to think of it a little. Most of us know that by injudicious training, horses, cattle, dogs, &c., may be rendered extremely vicious. If there is no perceptible analogy between these and bees, experience proves that they may be made ten times more irritable than they naturally would be.

THEIR MEANS OF DEFENCE.

Nature has armed them with means to defend their stores, and provided them with combativeness sufficient to use them when necessary. This could not be bettered. If they were powerless to repel an enemy, there are a thousand lazy depredators, man not excepted, who would prey upon the fruits of their industry, leaving them to starve. Had it been so arranged, this industrious insect would probably have long since been extinct.

TIME OF GREATEST IRRITABILITY.

The season of their greatest caution, in this section, is August, during the flowers of buckwheat. It is then their stores are greatest. As soon as a stock is pretty well supplied with this world's goods, like some bipeds, they become very haughty, proud, aristocratic, and insolent. A great many things are construed into insults, that in their days of adversity would pass unnoticed; but now it is becoming and proper for their honor to show a "just resentment." It behooves us, therefore, to ascertain what are considered insults.

PROPER CONDUCT.

First, all quick motions, such as running, striking, &c., about them, are noticed. If our movements among them are slow, cautious, humble, and respectful, we are often let to pass unmolested, having manifested a becoming deportment. Yet the exhalations from some persons appear very offensive, as they attack them much sooner than others; though I apprehend there is not so great a difference as many suppose. Whenever an attack is made, and a sting follows, the venom thus imparted to the air, if by only one, is perceived by others at some distance, which will immediately approach the scene, and more stings are likely to follow than if the first had not been.

HOW TO PROCEED WHEN ATTACKED.

Striking them down renders them ten times more furious. Not in the least daunted, they return to the attack. Not the least show of fear is perceived. Even after losing their sting, they obstinately refuse to desist. It is much the best way to walk as quietly as possible to the shelter of some bush, or to the house. They will seldom go inside of the door.

A PERSON'S BREATH OFFENSIVE, AND OTHER CAUSES.

The breath of a person inside the hive, or among them, when clustered outside, is considered in the tribunals of their insect wisdom as the greatest indignity. A sudden jar, sometimes made by carelessly turning up the hive, is another. After being once thoroughly irritated in this way, they remember it for weeks, and are continually on the alert; the moment the hive is touched, they are ready to salute a person's face. When slides of tin or zinc are used to cut off the communication between the hives and boxes, some of the bees are apt to be crushed or cut in two. This they remember, and retaliate, as occasion offers; and it may be when quietly walking in the apiary.

THEIR MANNER OF ATTACK.

I must disagree with any one who says we always have warning before being stung. I have been stung *a few times* myself. Two-thirds of them were received without the least notice—the first intimation was the "blow." At other times, when fully determined on vengeance, I have had them strike my hat and remain a moment endeavoring to effect their object. In this case, I have warning to hold down my face to protect it from the next attempt, which is quite sure to follow. As they fly horizontally, the face held in that position is not so liable to be attacked. When they are not so thoroughly charged with anger, they often approach in merely a threatening attitude, buzzing around very provokingly for several minutes in close proximity to our ears and face, apparently to ascertain our intentions. If nothing hostile or displeasing is perceived, they will generally leave; but should a quick motion or offensive breath offend them, the dreaded result is almost sure to follow. Too many people are apt to take these threatening manifestations as positive intentions to sting. When these things can be quietly endured, and at the same time leave their vicinity, it generally ends peaceably. They never make an attack while away from their home in quest of honey, or on their return, until they have entered the hive. It is only in the hive and its vicinity that we expect to meet this irascible temperament, which should not be allowed, or at least may be subdued in a great measure, if not entirely, by doing things in a quiet manner, and, by the use of tobacco smoke. Any person having the care of bees should go armed with this powerful weapon. As bees are not much affected with smoke, while flying in the air, but will have their own way, we must take them in the hive as the place to teach *them* a proper deportment!

Those who are accustomed to smoking will find a pipe or segar very convenient here. But such as are not would do better, perhaps, not to learn a bad habit. I will therefore give a simple substitute.

SMOKER DESCRIBED.

Get a tube of tin about five-eighths of an inch diameter, five or six inches in length; make stoppers of wood to fit both ends, two and a half or three inches long; with your nail-gimlet make a hole through them lengthwise: when put together it should be about ten inches. The ends may be tapered. On one end leave a notch, that it may be held with the teeth, which is the most convenient way, as you will often want to use both hands: it is also always ready, without any trouble to blow through, and also to keep the tobacco burning. When ready to operate, fill the tube with tobacco, ignite it, and put in the stoppers; by blowing through it you keep the tobacco burning while the smoke issues at the other end.

EFFECT OF TOBACCO SMOKE.

We can now subdue these combative propensities, or render them harmless; turn their anger to submission, and make them yield their treasures to the hands of the spoiler without an effort of resistance! When once overpowered, they seem to lose all knowledge of their strength, and no slave can be more submissive! After the effects of the smoke have passed off, their former animosity will return. Should any resentment be shown on raising a hive, blow in the smoke; they immediately retreat, "begging pardon." After a few times, they learn "it's no use," and allow an inspection. If you wish to take off a box, raise it just enough to blow under the smoke; there is no trouble; you can replace it with another; the bees are kept out of the way with a little more smoke, *and no anger created about it to be remembered.* Those in the box are all submission; they can be carried away and handled as you please, without a possibility of getting them irritated, until they once more get home, and then are much more "amiable" than if the box had been taken without the smoke. They seem to forget, or do not realize anything of the transaction. When bees are to be transferred to a new hive, it is unnecessary to be so very particular about the escape of a single bee; no fears need be entertained of such as get out. In driving, the loud humming indicates their submission; the upper hive can then be safely raised at any time. After being thus driven out, they may be pushed about with impunity, and still be quiet! In short, by using smoke on all occasions where they would be likely to be disturbed without it by our meddling with them, it has a tendency to keep dormant their combative propensities. When these have never been aroused, there is much less danger from their attacks while walking or looking among them. Any one wishing further proof, I would recommend the experiment of managing one year with smoke, and the next without.

STING DESCRIBED.

Their sting, as it appears to the naked eye, is but a tiny instrument of war; so small, indeed, that its wound would pass unheeded by all the larger animals, if it was not for the poison introduced at the same instant. It has been described as being "composed of three parts, a sheath and two darts. Both the darts are furnished with small points or barbs like a fishhook," that hold it when introduced into the flesh; the bee being compelled to leave it behind.

DOES ITS LOSS PROVE FATAL?

It is said "to the bee itself this mutilation proves fatal." This last is another assertion for fact, so often repeated, that perhaps we might as well admit it; seeing the difficulty we should have in disproving it. Only think of the impossibility of keeping our eye, for five minutes, on a bee that is flying about, after it has left its sting. Yet there are some persons so very particular about what they receive as facts, that they would require this very unreasonable thing of watching a bee till it died, before they could be *positively sure* that the loss of its sting caused its death. (It is much easier to guess.) They might even take analogy, and say that other insects possess so little sensation that they have been known to recover after much more extensive mutilation—that beetles have lived for months under circumstances that would have instantly killed some of the higher animals—that spiders often reproduce a leg, even lobsters can replace a lost claw, &c. I have put off describing any protection against their attacks, because I wish to get up a little more courage in our doings among them. Yet it is folly to expect all will manage successfully without something for defence.

MEANS OF PROTECTION.

The face and hands are most exposed; for the latter, thick woollen mittens or gloves are best; the sting is generally left when thrust into a leather glove. For the face procure one and a half yards of thin muslin or calico, sew the ends together, the upper end gathered on a string small enough to prevent it slipping over the head when put on. An arm-hole is to be cut out on each side; below is another string to gather it close to the body. As I do not expect you to work in the dark, we will have a place cut out in front, and a piece of coarse lace inserted; that which will just prevent a bee from passing, is best, as it gives us a better chance to see. To keep it from falling against the face, a wire is bent around and sewed fast. Any person that knows how to put on a shirt will manage this. When thus equipped, and other garments of proper thickness, the most timid ought not to hesitate to venture among them, when necessary. I cannot avoid cautioning you again to beware of irritating your bees, until this protection is necessary, as it is a rather bad state of things. With this on, you cannot conveniently use any smoke. To put this on and off is considerable trouble, and every time you go among them, if you have to resort to this, I fear some necessary duties will be neglected. Whenever a partial protection will do, I would recommend a handkerchief; it is always at hand, and can be put on in a moment; throw it over the head, letting the ends fall around the neck and shoulders, covering all but the face. The hat can come on over it. As for the face, whenever a bee comes around in a menacing attitude, hold it down—unless he stings at the first onset, there is not much risk.

REMEDIES FOR STINGS.

Concerning the remedies for stings, it is a hard matter to tell which is the best. There is so much difference in the effect in different individuals, and the different parts of the body, as well as the depth the sting reaches, that a great variety of remedies are recommended.

A person is slightly stung, and applies something as an antidote; the effect of the sting is trifling, as perhaps it would have been without anything, and the medicine is forthwith extolled as a sovereign remedy. I have been thus deceived; when slightly stung applied what I thought cured in one case, when in the next the sting might have penetrated deeper, or in some other place, and the remedy would seem to have no effect. For the last few years, I have not made any application whatever for myself, and the effect is no worse, nor even as bad as formerly. (This, I am told, is because the system is hardened, and now can resist or throw off the effects.) Among the remedies recommended, are saleratus and water, salt and water, soft-soap mixed with salt, a raw onion cut in two and one-half applied, mud or clay mixed pretty wet and changed often, tobacco wet and rubbed thoroughly to get at the strength, and cold water constantly applied. To cure the smart, the application of tobacco is strongly urged, and cold water is spoken of with equal favor to prevent the swelling.

When stung in the throat, drinking often of salt and water is said will prevent serious consequences.

Whether any of these remedies are applied or not, I suppose it is unnecessary to say that the sting should be pulled out as soon as practicable.

CHAPTER XIX.

ENEMIES OF BEES.

Among the enemies of bees, there are included rats, mice, birds, toads, and insects.

ARE THEY ALL GUILTY?

But some of these are probably clear of any actual mischief. I strongly suspect that the spirit of destructiveness with many people is altogether too active. There are some farmers, with this principle predominant, so short-sighted, that if it was in their power they would destroy a whole class of birds, because some of them had picked a few cherries, or dug out a few hills of corn, when, at the same time, they are indebted to their activity in devouring worms, insects, &c., that would otherwise have destroyed entire crops! It will be well, therefore, before condemnation, to see if on the whole we are to be gainers or losers by an indiscriminate slaughter, without judge or jury.

RATS AND MICE.

Rats and mice are never troublesome, except in cold weather. The entrances of all hives standing out are too small to admit a rat. It is only when in the house that much damage need be apprehended. They appear to be fond of honey, and when it is accessible will eat several pounds in a short time.

Mice will often enter the hive when standing on the bench, and make extensive depredations. Sometimes, after eating a space in the combs, they will there make their nest. The animal heat created by the bees will make a snug, warm place for winter quarters. There are two kinds: one the common class, belonging to the house; the other called "deer-mouse"—the under side perfectly white, the back much lighter than the other kind. The latter seems to be particularly fond of the bees, while the first appears to relish the honey. Whether they take bees that are alive, or only such as are already dead, I cannot say. Only a part of the bee is eaten; and if we take the fragments left to judge of the number consumed, the circumstance will go some ways to prove the sacrifice of quite a number. Whether bees or honey is wasted, a little care to prevent their depredations is well worthy of bestowal. As rats and mice have so long since been condemned and sentenced for being a universal plague, and without a redeeming trait, I will say nothing in their favor, and am perfectly willing they shall be hanged till dead.

ARE ALL THE BIRDS GUILTY?

But for some of the birds accused of preying upon bees, I would say a word.

KING-BIRD—ONE WORD IN HIS FAVOR.

The king-bird stands at the head of the list of depredators! With a fair trial he will be found guilty, though not so heinously criminal as many suppose. I think we shall find him guilty of taking only the drones. In the afternoon of a fair day he may be seen perched upon some dry branch of a shrub or tree near the apiary, watching for his victims, occasionally darting to seize them. I have shot him down and examined his crop, after seeing him devour a goodly number; but in every instance the bees were so crushed to pieces, that it was impossible to distinguish workers from drones. We are told of great numbers of workers being counted. It may be so, or it may be thus represented by a spice of prejudice. I have found the brutal gratification of taking life so strong with some, that a natural antipathy is allowed to take the place of justice, and a proper defence is not allowed in such cases where the suffering party has not the power to enforce it. If he was satisfied with workers as well as drones, why does he not visit the apiary long before noon, and fill his crop with them? But instead, he waits till afternoon for the drones; and if none are flying, he watches quietly till one appears, although workers may be out by hundreds continually. If the question is asked, how they tell the difference in the two kinds of bees, I might suggest that *instinct* has taught most animals the proper kind of food, and might direct the birds in this case. If it was not sufficient, a little experience in catching bees provided with stings, might impart the important difference, in one or two lessons. I once had a chicken that knew the difference by some means, and would stand by the hive and devour every drone, the moment it touched the board, while the workers would pass by him in scores untouched!

Now, whether this taking the drones is a disadvantage or otherwise, would depend entirely upon circumstances. If honey was a little scarce, the less we had of them the better; it would also save the bees some trouble in dispatching them. It is probably a matter of so little moment to our bees, that it will not pay for powder to shoot them.

Martins, and a kind of swallows, are said to be guilty of taking bees on some occasions; but as they pursue them on the wing (if they do), the same remarks will apply as to the king-bird.

CAT-BIRD ACQUITTED.

The cat-bird also comes in for a share of censure. It is said "they will get right down by the hive, and pick up bees by the hundred." Yet, right in the face of this charge, I am disposed to acquit him. With the closest observation, I find him about the hive, picking up *only* young and immature bees, such as are removed from the combs and thrown out. They may be seen as soon as the first rays of light make objects visible about the apiary, looking for their morning supply, as well as frequent visits during the day. Should an unlucky worm be in sight just then, while looking up a place for spinning a cocoon, or a moth reposing on some corner of the hive, their fate is at once decided. Before destroying this bird, it would be well to judge by actual observation as to facts; otherwise we might "destroy a friend instead of a foe."

TOAD GOT CLEAR.

A toad is discovered near the hives, and forthwith he is executed as a bee-eater. "He ought to be killed for his looks, if nothing else!" He is thus often sacrificed *really* on account of his appearance, while pretending he is a villain. It is true his "feathers" will not vie in brilliancy with the plumage of the humming-bird, and do not gratify ideality—therefore he is dispatched. The next week the complaint is made that the little bugs, that he might have destroyed, "have eaten up all the little cucumbers and cabbages." His food is probably small insects. Whoever has seen him swallow bees, must have watched closer than I ever did.

WASPS AND HORNETS NOT FAVORED.

As for the frequent visits of the black-wasp in the sunny days of spring, but little can be said in their favor—they seem to have no other object but to tease and irritate the bees. I never could discover that they entered the hive for the purpose of plunder. They have frequent battles with the bees, but I never saw any bees devoured or carried off, nor even killed. After the first of June they are seldom troublesome. The yellow wasp or hornet, that is around in autumn, is of but little account; their object is honey, which they take when they can get it, but are not apt to enter the hive among the bees.

ANTS—A WORD IN THEIR FAVOR.

Ants come in for a share of condemnation. This little industrious insect shall have my endeavors for a fair hearing; I think I can understand why they are so frequently accused of robbing bees. Many bee-keepers are wholly ignorant, most of the time, of the real condition of their stocks. Many causes independent of ants, induce a reduction of population. Suppose the bees are so reduced as to leave the combs unprotected, and the ants enter and appropriate some of the honey to themselves, and should the owner come along just then and see them engaged, "Ha! you are the rascals that have destroyed my bees," without a thought of looking for causes, beyond present appearances. They are often unjustly accused by the farmer of injuring the growth of his little trees, by causing the tender leaves to curl and wither. Inquiries are often made in some of the agricultural papers for means to destroy them, merely because they are found on them; when the real cause of the mischief is with the plant louse, (aphis) that is upon the leaves or stalk in hundreds, robbing them of their important juices, and secreting a fluid greatly prized by the ants. By destroying the lice, you remove all the attraction of the ants. The peculiar habits of the small black ants, probably give rise to a suspicion of mischief in this way. They live in communities of thousands—their nests are usually in old walls, in old timber, under stones, and in the earth. From their nests a string may be traced sometimes for rods, going after, and returning laden with food. During a spell of wet weather, such as would make the earth and many other places too damp and cold for a nest, they look out for better quarters. The top or chamber of our bee-hives affords shelter from rain. The animal heat from the bees renders it perfectly comfortable. How then can we blame them for choosing such a location, so completely answering all their wants? As long as the bees are not disturbed, we can put up with it better. But the careless observer having discovered their train to and fro from their nest on the hive, exclaims: "Why, I have seen them going in a continual stream to the hive after honey;" when a little scrutiny into the matter would show that only the nest was on the top of the hive, and they were going somewhere else for food; not one to be seen entering the hive among the bees for honey, (at least I never could detect it.)

When honey is unprotected by bees, or boxes of it placed where they can have access, as a natural consequence, they will carry off some; but it is easily secured.

SPIDER CONDEMNED.

Spiders are a source of considerable annoyance to the apiarian, as well as to the bees; not so much on account of the number of bees consumed, as their habit of spinning a web about the hive, that will occasionally take a moth, and will probably entangle fifty bees the whilst. They are either in fear of the bees, or they are not relished as food; particularly, as a bee caught in the morning is frequently untouched during the day. This web is often exactly before the entrance, entangling the bees as they go out and return; irritating and hindering them considerably. They often escape after repeated struggles. I have removed a web from the same place every morning, for a week, that was renewed at night with astonishing perseverance! I can generally look out his hiding-place, which is in some corner near by, and dispatch him. His redeeming qualities are few, and are more than balanced by the evil, as far as I have discovered. Their sagacity in some instances will find a place of concealment not easily discovered. At the approach of cold weather, the box or chamber of the hive being a little warmer than other places, will attract a great many there to deposit their eggs. Little piles of webbing or silk may be seen attached to the top of the hive, or sides of boxes. These contain eggs for the next year's brood. This is the time to destroy them and save trouble for the future.

If we combine into one phalanx all the depredators yet named, and compare their ability for mischief with the wax moth, we shall find their powers of destruction but a small item! Of the moth itself we would have nothing to fear were it not for her progeny, that consist of a hundred or a thousand vile worms, whose food is principally wax or comb.

As the instinct of the flesh-fly directs her to a putrid carcass to deposit her eggs, that her offspring may have their proper food, so the moth seeks the hive containing combs, and where its natural food is at hand to furnish a supply. During the day a rusty brown miller, with its wings wrapped close around the body, may be often seen lying perfectly motionless on the side of the hive on one corner, or the under edge of the top, where it projects over— they are more frequent at the corners than anywhere else, one-third of their length projecting beyond it; appearing much like a sliver on the edge of a board that is somewhat weather-beaten. Their color so closely resembles old wood, that I have no doubt their enemies are often deceived, and let them escape with their lives. As soon as daylight shuts out the view, and no danger of their movements being discovered by their enemies, they throw off their inactivity, and commence searching for a place to deposit their eggs, and woe to the stock that has not bees sufficient to drive them from the comb. Although their larvæ has a skin that the bee cannot pierce with its sting, in most cases, it is not so with the moth, and of this fact they seem to be aware, for whenever a bee approaches they dart away with speed ten times greater than that of any bee, disposed to follow! They enter the hive and dodge out in a moment, having either encountered a bee, or fear they may do so. Now it needs no argument to prove that when all our stocks are well protected, that it must be a poor chance to deposit eggs, on the combs of such hives, where their instinct has taught them is the proper place. But they *must* leave them somewhere. When driven from all the combs within, the next best place is the cracks and flaws about the hive, that are lined with propolis; and the dust and chips that fall on the floor-board of a young swarm not full will be used. This last material is mostly wax, and answers very well instead of comb. The eggs will here hatch and the worms sometimes ascend to the combs; hence the necessity of keeping the bottom brushed off clean. It will prevent those that are on the bottom from going up; also the bees from taking up any eggs, if this should happen to be the method. I can conceive of no other way by which they get among the combs of a populous stock; where they are often detected, having been deposited by some means. A worm lodged in the comb, makes his way to the centre, and then eats a passage as he proceeds, lining it with a shroud of silk, gradually enlarging it, as he increases in size. (When combs are filled with honey, they work on the surface, eating only the sealing.) In very weak families this silken passageway is left untouched,—but removed by all the stronger ones. I have found it asserted that "the worms would be all immediately destroyed by the bees, were it not for a kind of dread in touching them until compelled to by necessity." As the facts which led to this conclusion are not given, and I can find none confirming it, perhaps I shall be excused if I have no faith. On the contrary, I find to all appearance an instinctive antipathy to all such intruders, and are removed immediately when possessing the power.

When a worm is in a comb filled with brood, its passage being in the centre, it is not at first discovered. The bees, to get it out, must bite away half the thickness, removing the brood in one or two rows of cells, sometimes for several inches. This will account for so many immature bees found on the bottom board at morning, in the spring; as well as in stocks and swarms but partially protected after the swarming season.

INDICATIONS OF THEIR PRESENCE.

Sometimes a half dozen young bees, nearly mature, will be removed alive, all webbed together, fastened by legs, wings, &c. All their efforts for breaking loose prove unavailing. Also others that are separate may be seen running about with their wings mutilated, or part of their legs eaten off, or tied together! These generally are the first symptoms of worms in our stock at this season. Although unfavorable, it might be worse. It shows that the bees are not discouraged yet,—that when finding the worms present, have sufficient energy left to make an effort to rid themselves of the nuisance.

MANAGEMENT.

Should the apiarian now give them a little assistance for a few days, they will soon be in a prosperous condition. The hive should be frequently raised, and everything brushed out clean. If it is a new swarm half full, that presents these indications, it should be turned over a few times, perhaps once a week, till the worms are mastered; and the corners below the bees examined for the cocoons, that will very often be found there, and are easily detached and destroyed. In turning over a hive part full, in warm weather, you should first observe the position of the combs, and let the edges rest against the side of the hive, otherwise they might bend over and break loose when the hive was again set up, (by simply making a pencil mark across the top in the direction of the combs, you may know any time after first looking).

CARE IN TURNING OVER HIVES.

When a hive is full of combs, the edges are usually attached sufficient to steady them, and it is of less consequence which way it is turned, yet in very warm weather the honey will run out of drone cells if perpendicular.

In *very* small swarms, hundreds of the young brood may be frequently seen with their heads out of the cells, endeavoring to escape, but are firmly held inside by these webs. I have known a few instances in such circumstances, where it appeared as if the bees had cut off the whole sheet of comb and let it drop, thereby ridding themselves of all further trouble (or would be rid of it, if their owner only did his part by taking out what fell down.)

OTHER SYMPTOMS OF WORMS.

But when the bees make no effort to dislodge the enemy or his works in old stocks, the case is somewhat desperate! Instead of the foregoing symptoms we must look for something entirely different. But few young bees will be found. In their place we may find the fæces of the worms dropped on the board. During winter and spring the bees, in biting off the covering of cells to get at the honey, drop chips closely resembling it. To detect the difference and distinguish one from the other requires a little close inspection. The color of the fæces varies with the comb on which they feed, from white to brown and black. The size of these grains will be in proportion to the worm—from a mere speck to nearly as large as a pin-head: shape cylindrical, with obtuse ends: length about twice its diameter. By the quantity we can judge of the number. If the hive is full of combs the lower ends may appear perfect, while the middle or upper part is sometimes a mat of webs!

Whenever our stocks have become reduced from over-swarming or other cause, this is the next effect in succession that we must expect. Here is another important reason that we know the *actual* condition of our bees at all times; we can then detect the worms very soon after they commence. In some instances we might save the stock by breaking out most of the combs, leaving just enough to be covered by the bees. When success attends this operation, it *must* be done before the worms have progressed to a thorough lodgment. When the stock is weak, and appearances indicate the presence of many, it is generally the safest, and will be the least trouble in the end, to drive out the bees at once and secure the honey and wax. The bees when put into a new hive *may* do a little, but if they should do nothing, it would be no worse. It cannot be as bad any way as to have left them in the old hive till the worms had destroyed all and matured a thousand or two moths in addition to those otherwise produced, thereby multiplying the chances of damage to other stocks a thousand-fold. It is probably remembered that I said when bees are removed from a hive in warm weather, if it was not infested with worms at the time, it soon would be, unless smoked with sulphur.

WHEN THEY GROW LARGER THAN USUAL.

In a hive thus left without bees to interfere, the worms will increase to one-half or two-thirds larger than where their right to the combs is disputed. In one case they often have their growth, and actually wind up in their cocoon when less than an inch in length: in the other they will quietly fatten till they are an inch and a half long and as large as a pipe-stem.

TIME OF GROWTH.

When first hatched from the egg, it requires very close inspection to see them with the naked eye. The rapidity of growth depends on the temperature in which they are, as much or more than their good living. A few days in hot weather might develop the full-grown worm, while in a lower temperature it would require weeks and even months in some cases, perhaps from fall till spring.

TIME OF TRANSFORMATION.

The worm, after spinning its cocoon, soon changes to a chrysalis, and remains inactive for several days, when it makes an opening in one end and crawls out. The time taken for this transformation is also governed by the temperature, although I think but few ever pass the winter in this state. It is a rare thing to find a moth before the end of May, and not many till the middle of June; but after this time they are more numerous till the end of the season.

FREEZING DESTROYS WORMS, COCOON AND MOTH.

It is pretty well demonstrated that the moth, its eggs, larvæ and chrysalis cannot pass the winter without warmth of some kind to prevent their freezing to death. The following facts indicate this. I have taken all the bees out of a hive in the fall, and without disturbing the comb or honey, put it in a cold chamber where it could freeze thoroughly. In the following March bees were again introduced, and when not on a bench with some other stock that had worms, not a single instance in forty cases has ever produced a worm before the middle of June, or until the eggs of some moth matured in another hive has had time to hatch. I have sometimes, instead of putting bees in these in March, kept them till June for swarms, perfectly free from any appearance of worms!

HOW THEY PASS THE WINTER.

But it is altogether a different thing with our hives in which bees are wintered; they are seldom or never entirely exempt! Perhaps it is impossible to winter bees without preserving some eggs of the moth or a few worms at the same time. The perfect moth perhaps never survives the winter; the only place that the chrysalis would be safe, I think must be in the vicinity of the bees—and a good stock will never allow it there—but eggs, it would appear, are suffered to remain. In the fall, at the approach of cold weather, the bees are apt to leave the ends of the combs exposed; the moth can now enter and deposit her eggs directly upon them; these, together with what are carried in by means before suggested, are enough to prevent losing the breed. The warmth generated by the bees will keep these eggs from freezing and preserve their vitality. When warm weather approaches in the spring, those nearest the bees are probably hatched first, and commence depredations and are removed by the bees. As the bees increase and occupy more comb, more are warmed up and hatched. In this way, even a small family of bees will hatch, and get rid of all the eggs that happen to be in their combs, and not be destroyed. This is the time that the apiarian may be of service in destroying the worms, as the bees get them on the floor.

STOCKS MORE LIABLE TO BE DESTROYED LAST OF SUMMER.

But in July and August it is different in this respect; a single moth may enter the hive when exposed, and deposit her whole burden of several hundred eggs, as in the other case, but the heat from the bees is now unnecessary to hatch them. The weather at this season will make any part of the hive warm enough to set her whole brood at work at once, and in three weeks all may be destroyed! This, and the fact that more moths exist now than before, may account for the greater number of stocks being destroyed at this season. Yet it is considered extremely bad management to allow honey or combs to be devoured by this disgusting creature. A little care to know the condition of the stocks *is necessary* to prevent their getting the start. These duties should be fully considered before we take the responsibility of the care of bees.

WHEN BEES ARE SAFE.

The only condition when we can rest and feel safe is when *we know all our stocks are full of bees.* Even the "moth-proof" hive containing combs will be scented out by the moth, when there are no bees to guard it. An argument to show that a moth can enter where a bee can go is unnecessary, and a little observation, I think, will prove that her eggs sometimes go where she is not allowed.

MEANS TO DESTROY THEM.

At this season, (July and August), it is a good plan to put a few pieces of old dry combs near the hives, in a box, or other place, as a decoy, where the moth may have access. She will deposit a great many of her eggs here, instead of the hive, and can be easily destroyed. As we cannot always have our bees in a situation to feel safe, it will be well to adopt some of the means recommended to diminish the number of moths. First destroy all the worms that can be found at any time, particularly in spring; second, all cocoons that can be got at. A great many worms can be enticed to web up, under a trap of elder, &c., when it is an easy matter to dispatch them. Thirdly, destroy all the moths possible that can be seen about the hive. They are very much like the flea, "when you put your finger on him he is not there;" a careful move is necessary to crush him at once, otherwise he darts away at the least disturbance. Probably the most expeditious mode is to make them drunk.

MAKING THEM DRUNK, AND THEIR EXECUTION BY CHICKENS.

Mix with water just enough molasses and vinegar to make it palatable; this is to be put in white saucers or other dishes, and set among the hives at night. Like nobler beings, if not wiser, when once they have tasted the fatal beverage, they seem to lose all power to leave the fascinating cup; but give way to appetite and excitement till a fatal step plunges them into destruction! The next morning finds them yet wallowing in filth, weak and feeble. Whether they would recover from the effects of their carousal if lifted out of the mire, and carefully nursed like other specimens of creation, I never ascertained. With but little trouble a chicken or two will learn to be on hand, and greedily devour every one. Hundreds are caught in this way, although many other kinds besides the bee-moth will be mixed with them. This drink may be used till dried up, occasionally adding a little water; perhaps it is better after fermenting. This recipe appeared some years ago in some paper; I have forgotten where. Salt has been recommended to prevent the mischief of the worms, as well as a benefit to the bees. I used it pretty extensively for several years, as I thought without much benefit, and got tired. I then tried salting a part, and let the rest do entirely without, and found no difference in their prosperity. Since then, some ten years ago, I abandoned its use altogether, and succeed just as well.

CHAPTER XX.

MELTING DOWN OF COMBS.

THE CAUSE.

When extreme hot weather occurs immediately after the bees have been gathering from a plentiful harvest for two or three weeks, or even during the yield, the wax composing new combs is very liable to be softened, till they break loose from their fastenings and settle to the bottom.

EFFECTS.

Sometimes the injury is trifling, only a piece or two slipping down; at other times the whole contents fall in a confused and broken mass, the weight pressing out the honey, and besmearing the bees, which in that situation creep out, and away, from the hive in every direction.

I once had some new stocks ruined, and several others injured by hot weather, in this way, about the first of September, immediately after the flowers of buckwheat. The bees, or most of them, being covered with honey, together with what ran out of the hive, at once attracted bees from the others to the spot, which carried off the entire contents in a few hours. This was an uncommon occurrence; I have known but one season in twenty-five years when it occurred after the failure of honey in the flowers. It usually happens during a plenteous yield, and then other stocks are not apt to be troublesome.

FIRST INDICATIONS.

The first indications of such an accident will be, the bees outside in clusters, when the hive is perhaps only half or two-thirds full, and the honey running out from the bottom, (this is when part has fallen.)

PREVENTION.

To prevent such occurrences as far as possible, ventilate by raising the hives on little blocks at the corners, and *effectually protect them from the sun*; and if necessary, wet the outside with *cold* water. At the time of losing those before mentioned, I kept all the rest of the young swarms wet through the middle of the day, and I have no doubt but I saved several by this means. I had some trouble with such as had only a piece or two come down, and started just honey enough to attract other bees. It was not safe to close the hive to prevent the robbers, as this would have made the heat still greater, and been certain destruction.

The best protection I found, was to put around the bottom of the hive a few stems of asparagus; this gave a free circulation of air, and at the same time, made it very difficult for the robbers to approach the entrance, without first creeping through this hedge and encountering some bees that belonged to the hive; which, with this assistance, were enabled to defend themselves till all wasting honey was taken up.

When the hive is nearly full, and but one or two sheets come down, the lower edge will rest on the floor, and the other combs will keep it in an upright position, until the bees fasten it again. It is generally as well to leave such pieces as they are. If the hive is but half full or little more, and such pieces are not kept perpendicular by the remaining combs, they are apt to be broken and crushed badly, by falling so far; and most of the honey will be wasted. To save this, it will be necessary to remove it, (unless a dish can be made to catch it). Be careful not to turn the hive on its side, and break the remaining combs, if any are left. Such combs as contain brood and but little honey, might be left for the brood to mature. Should the bees be able to take the honey or not waste much, it might be advisable to leave it, till the contents were taken up; it would greatly assist in filling up. But these broken pieces should be removed before they interfere with the combs extending to the bottom. A part of the bees are generally destroyed, but the majority will escape; even such as are covered with honey, (if they are not crushed) will clean it off and soon be in working order, when others do not interfere officiously, assisting to remove it. A good yield of honey is the best protection against this disposition to pillage. After the first year combs become thicker, and are not so liable to give way.

CHAPTER XXI.

FALL MANAGEMENT.

FIRST CARE.

When the flowers fail at the end of the season, the first thing necessary is to ascertain which are the weakest stocks, and all that cannot defend themselves should either be removed or reinforced. The strength of all stocks is pretty thoroughly tested within a few days after a failure of honey. Should any be found with too few bees for defence, they are quite sure to be plundered. Hence the necessity of action in season, that we may secure the contents in advance of the robbers.

STRONG STOCKS DISPOSED TO PLUNDER.

Strong stocks, that during a yield have occupied every cell with brood and honey, when it fails, will soon have empty cells left by the young bees, hatching. These empty cells, without honey to fill them, appear to be a source of much uneasiness. Although such hive and caps may be well stored, I have ever found them to be the worst in the apiary, much more disposed to plunder, than weaker ones with half the honey. As weak stocks cannot be bettered now, it is best to remove them at once, and put the temptation out of the way. Carelessness is but a sorry excuse, for letting bees establish this habit of dishonesty. Should any stocks be weak from disease, the consequences would be even more disastrous than bad habits; the reasons why such impure honey should not go into thrifty stocks, have already been given. If we want the least possible trouble with our bees, none but the best should be selected for winter. But what constitutes a good stock, seems to be but partially understood; if we judge from the number lost annually, too many are careless, or ignorant in the selection; supposing, perhaps, because a stock has been good one winter and swarmed well, it must of course be right; the mistake is often fatal.

BEES CHANGEABLE.

Bees are so changeable, especially in the summer and swarming season, that we can seldom be certain what they are, by what they have been. It is safest, therefore, *to know what they are now*.

REQUISITES FOR GOOD STOCKS.

The proper requisites for a good stock are a full hive of proper shape and size, (viz., 2,000 inches,) well stored with honey; a large family of bees, and in a healthy condition, which must be ascertained by actual inspection. The age is not important till over eight years old. Stocks possessing these points, can be wintered with but little trouble. But it cannot be expected that all will be in this condition. Many bee-keepers will wish to increase their stocks and keep all that is practicable, by supplying any deficiency. I shall endeavor to make it appear profitable to do so, until bees enough are kept in the country, to get all the honey that is now wasted.

All can understand why it is a loss to have bees eat honey part of the winter and then die—that the honey consumed might have been saved—that it makes no great difference to the bees whether they are killed in the fall or sacrificed in the winter. I am not an advocate for fire and brimstone as the reward of all unfortunate stocks, and shall recommend it only when its use will make it no worse. We will see how far it can be dispensed with.

GREAT DISADVANTAGE OF KILLING THE BEES.

Those rustic bee-keepers who are in the habit of making their hives very large, such as will hold from 100 to 140 lbs., and killing the bees in the fall, and sending the honey to market, will probably continue the use of sulphur, unless we can convince them of the greater advantage of making the hive smaller and have fifty or eighty lbs. of this honey in boxes which will sell for more than can be realized for their larger hive full, and at the same time, save their bees for a stock-hive, making a better return in the long run, than one hundred dollars at interest. When hives are made the proper size, the honey will not be an object sufficient to pay for destroying the bees.

SECTION OF COUNTRY MAY MAKE A DIFFERENCE IN WHAT POOR STOCKS NEED.

The kind of requisite to be supplied to our deficient stocks, will probably depend on the section of country. Where the principal source is clover and basswood, it will fail partially, at least, before the end of warm weather.

Some poor or medium stocks will continue to rear brood too extensively for their means, and exhaust their winter stores in consequence; such will need a supply of honey. But where great quantities of buckwheat are sown, cold weather follows almost immediately after this yield, and stops the breeding. Consequently a scarcity of bees is more frequent than honey. There are exceptions, of course; I am speaking of these cases generally. My experience has mostly been in a section where this crop is raised, and will say that there is not more than one season in ten, but that the honey will be in proportion with the bees the first of September; that is, if there are bees enough, there will be honey enough.

WHEN BEES ARE NEEDED.

I have frequently had stocks with stores amply sufficient to carry a good family through the winter, and yet too few bees to last till January, or even to defend themselves from the robbers. Hence I am in the habit of supplying bees oftener than honey.

I usually have some few hives with too little honey, as well as too few bees. Now it is very plain if the bees of one or more of this class were united with the first successfully, we should have a respectable family. I have made additions to stocks in this way that proved first-rate.

CAUTION.

Whenever we make additions in this manner, it would be well first to ascertain what was the cause of a scarcity of bees; if it was over-swarming or loss of queen, it is well enough—but if from disease, reject them, unless the bees are to be transferred the next spring, and then, when too many cells are occupied with dead brood, as the bees cannot be successfully wintered.

PRINCIPAL DIFFICULTY.

The greatest difficulty in uniting two families or more in this manner, is where they have to be taken from different places in the same apiary; where the locations have been marked. It is sufficiently shown that bees return to the old stand.

To prevent these results, it has been recommended "to set an empty hive with some pieces of comb, fastened in the top in the place of the one removed, to catch the bees that go back to the old stand, and remove them at night for a few times, when they remain." This should be done only when we cannot do better; it is considerable trouble; besides this, we do not always succeed to our satisfaction.

HOW AVOIDED.

I like the plan of bringing them a mile or more for this purpose, and have no after trouble about it. Two neighbors being that distance apart, each having stocks in this condition might exchange bees, making the benefit mutual. I have done so, and considered myself well paid for the trouble. But latterly I have had several apiaries away from home, and now manage without difficulty.

ADVANTAGES OF MAKING ONE GOOD STOCK FROM TWO POOR ONES.

This making one good stock out of two poor ones, cannot be too highly recommended; aside from its advantages, it relieves us from all disagreeable feelings in taking life, that we can with but little trouble preserve.

TWO FAMILIES TOGETHER WILL NOT CONSUME AS MUCH AS IF SEPARATE.

Even when a stock already contains bees enough to make it safe for winter, another of the same number of bees may be added, and *the consumption of honey will not be five lbs. more than one swarm would consume alone*. If they should be wintered in the cold, the difference might not be one pound. Why more bees do not consume a proportionate quantity of honey, (which the experience of others as well as myself has thoroughly proved), is a mystery, unless the greater number of bees creates more animal heat, and being warm, eat less, is a solution, (which if it is, is a strong reason for keeping bees warm in winter.)

AN EXPERIMENT.

Notwithstanding all this, I cannot recommend making a *good* stock better by adding the bees from another good one as a source of profit. I tried it a few times. I had purchased some large hives for market, and wished to dispose of the bees without sulphur, and try the experiment of uniting two or more. The next spring when they commenced work such double stocks promised much; but when the swarming season arrived, the single swarms, such as were good and had just about bees enough, were in the best condition, in ordinary seasons. Whether this was owing to the circumstance of there being already bees enough that were beginning to crowd and interfere with each other's labors, and less brood raised in consequence, or to some other reason, I cannot say. I have often noticed, (as others have), that stocks which have cast no swarms, are no better the next spring than others. The same cause might operate in both cases. Therefore it would appear unnecessary to unite two or more *good swarms*, unless it is to spare our feelings in destroying the bees. The two extremes may generally be avoided, and not have too many or too few bees.

SEASON FOR OPERATING.

The season for operating is, generally, when all the brood has matured and left the cells. The exceptions are when there are not bees enough to protect the stores; it may then be necessary, immediately after the failure of honey.

Col. H. K. Oliver, of Salem, Mass., is said to be the inventor of the fumigator, an instrument to burn fungus (*puff-ball*). By the aid of this the smoke is blown in the hive, paralyzing the bees in a few minutes; when they fall to the bottom, apparently dead, but will recover in a few minutes, on receiving fresh air.

THE FUMIGATOR.

I am indebted to a communication from J. M. Weeks, published on page 151 of the Cultivator for 1841, for this method. The description of the fumigator that I constructed will vary a trifle from his, but will retain the principle. I obtained a tin tube four inches long, and two in diameter. Next, I made a stopper of soft wood, three inches long, to exactly fit one end of the tube when driven in half an inch, and secured it by little nails driven through the tin. Through the centre of this stopper I made a hole one-fourth of an inch in diameter. To prevent this hole filling up, the end in the tube was covered over with wire cloth, made a little convex. The end of this stopper was cut down to about half an inch, tapering it from the tin. For the other end a similar piece of wood is fitted, though a little longer, and not to be fastened, as it must be taken out for every operation. The outer end of this is cut down into a shape to be taken into the mouth, or attached to the pipe of a bellows. (I fitted them in the turning lathe, but have seen them fixed very nicely without.) It could all be made of tin; but then it is necessary to use solder, which is liable to melt and cause leaks.

FUMIGATOR.

"The puff-balls must not be too much injured by remaining in the weather, and should be picked, if possible, just before they are ripe and burst open. When not thoroughly dry, put them in the oven after the bread is out." When used, the cuticle or rind must be carefully removed; ignite it by a lamp or coal (it will not blaze in burning), blow it, and get it thoroughly started, before putting it in the tube. Put in the stopper, and blow through it; if it smokes well, you are ready to proceed. When it does not burn freely, unstop and shake it out. The dry air is much better than moist breath at the commencement.

DIRECTIONS FOR UNITING TWO FAMILIES.

The hive to receive the bees is inverted, the other set over it right end up, all crevices stopped to prevent the escape of the smoke. Now insert the end of the fumigator into a hole in the side of the hive (which if not made before will need to be now); blow into the other end, this forces the smoke into the hive; in two minutes you may hear the bees begin to fall. Both hives should be smoked; the upper one the most, as we want all the bees out of that. The other only needs enough to make the scent of the bees similar to those introduced. At the end of eight or ten minutes, the upper hive may be raised, and any bees sticking between the combs brushed down with a quill. The two queens in this case are of course together; one will be destroyed, and no difficulty arise. But if either of them is a young one, and you have been convinced by some "bee-doctor" that such are much more prolific, and happen to know which hive contains her, and wish that one to be preserved, you can do so by varying the process a little. Instead of inverting one hive, set them both on a cloth right side up, and smoke the bees; the queens are easily found, while they are all paralyzed; then put the bees all together. The hive should now have a thin cloth tied over the bottom, to prevent the escape of the bees. Before they are fully recovered, they seem rather bewildered, and some of them get away. Set the hive right end up, and raise it an inch; the bees drop on the cloth, and fresh air passing under soon revives them. In from twelve to twenty-four hours, they may be let out.

Families put together in this way will seldom quarrel (not more than one in twenty), but remain together, defending themselves against intruders as one swarm.

I once had a stock nearly destitute of bees, with abundant stores for wintering a large family. I had let it down on the floor-board, and was on the lookout for an attack. The other bees soon discovered this weakness, and commenced carrying off the honey. I had brought home a swarm to reinforce them only the day before, and immediately united them by means of the fumigator. The next morning I let them out, allowing them to issue only at the hole in the side of the hive. It was amusing to witness the apparent consternation of the robbers that were on hand for more plunder; they had been there only the day before, and had been allowed to enter and depart without even being questioned. But lo! a change had come over the matter. Instead of open doors and a free passage, the first bee that touched the hive was seized and very rudely handled, and at last dispatched with a sting. A few others receiving similar treatment, they began to exercise a little caution, then tried to find admission on the back side, and other places; and attempted one or two others on either side, perhaps thinking they were mistaken in the hive; but these being strong, repulsed them, and they finally gave it up. I mention this to show how easy it is, with a little care, to prevent robberies at this season. Too many complaints are made about bees being robbed; it is very disagreeable. Suppose that *none were plundered through carelessness*; this complaint would soon be a rare thing.

UNITING WITH TOBACCO SMOKE.

By the use of tobacco smoke, bees may be united with nearly the same success. First, smoke the two to be united, thoroughly; disturb them and smoke again, that all may become partially drunk, and acquire the same scent. Then invert both hives, and with your pruning tools, cut the combs down on the sides of the hive, and across the top, and take out one comb at a time with the bees on it, and brush them with a quill into the other hive; they immediately go down among the combs, without once thinking it necessary to sting you. When done, the bees are to be confined, the same as in the other method. I do not like this method as well as the first, and do not resort to it when I can get the puff-ball. The bees are more liable to disagree, and it compels me to take out the comb, which I do not always like to do at the time. To avoid it, I have tried to drive them, but when the hive is only part full of combs, or contains but few bees, it is a slow job; and more so in cool weather.

CONDITION OF STOCKS IN 1851.

The latter part of the summer of 1851 was very dry and cold; the yield of buckwheat honey was not a tenth of the usual quantity; the consequence was, that none but early swarms had sufficient honey for winter; twenty-five pounds is required to make it *safe* in this section. I had over thirty young swarms with less than that quantity. Feeding for winter I avoid when I can; they would not winter as they were; and yet I made the most of them good stocks for the next summer by the following plan.

HOW THEY WERE MANAGED.

I had about twenty old stocks with diseased brood, and but few bees, yet *honey enough*. Now this honey appears healthy enough for the old bees, and fatal only to the young brood.

I transferred the bees of these new swarms to the old stocks with black comb and diseased brood. The bees were thus wintered on honey of but little account any way, and all that was in the others, new and healthy, was saved. These new hives were set in a cold dry place for winter; *right end up*, to prevent much of the honey from dripping out of the cells; some will leak then, but not as much as when the hive is bottom up. Honey that runs out, when the hive is bottom up, will soak into the wood at the base of the combs; this will have a tendency to loosen the fastenings, and render them liable to fall, &c.

The next March the bees were again transferred from the old to the new hives. My method is as follows: As the combs in the hive to receive the bees are rather cold, I set them by the fire, or in a warm room, for several hours previous. I take a warm room before a window, and as some few bees fly off, they will collect there. The new hive is turned bottom up on the floor; the old one on a bench by the side of it, having smoked the bees to keep them quiet. One comb at a time is taken out, and the bees brushed into the new hive; (a little smoke will keep them there). When through, I get the few on the window, and tie over a cloth to confine them, and keep them warm for a few hours longer. Paralyzing with puff-ball will answer instead, but they do not always all fall out of the combs when the hive is filled to the bottom, and it is possible that if a few were left, the queen might be one. Also a very few bees are worth saving at this season, and the combs might have to be broken out at last, for this purpose.

When a good-sized family is put in a hive containing fifteen or twenty pounds of honey, and near half full of clean new comb, they are about as sure to fill up and cast a swarm, as another that is full and has wintered a swarm.

CAUSE OF THEIR SUPERIOR THRIFT.

One cause of superior thrift may be found in the circumstance, that all moth eggs and worms are frozen to death, and the bees are not troubled with a single worm before June. No young bees have to be removed to work them out. Nearly every young bee that is fed and sealed up, comes forth perfect, and of course makes a vast difference in the increase.

SWARMS PARTLY FILLED PAY BETTER THAN TO CUT OUT THE HONEY.

Any person wishing to increase his stocks to the utmost, will find this plan of saving all part-filled hives, of much more advantage than to break it out for sale. Suppose you have an old stock that needs pruning, and have neglected it, or it has refused to swarm, and give you a chance without destroying too much brood. You can let it be, and put on the boxes; perhaps get twenty-five pounds of cap honey; and then winter the bees as described, and in the spring transfer them to the new combs. Again, if there is no stocks to be transferred in the spring, keep them till the swarming season. If a swarm put into an empty hive would just fill it, the same swarm put into one containing fifteen pounds of honey, it seems plain, would make that number of pounds in boxes. The advantage is, in the comparative value of box or cap honey over that stored in the hive; the difference being from thirty to a hundred per cent.

ADVANTAGES IN TRANSFERRING.

I would now like to show the advantages I derived in transferring the twenty swarms before mentioned. We will suppose that each family, from the first of October till April, consumed twenty pounds of honey. That in the centre combs, where there is most bee-bread, &c., is eaten first; if any is left, it is at the top and outside. If I had attempted to take out and strain this twenty pounds in the fall, it would have been so mixed with dead brood, and bee-bread, that I probably should have rejected most of it. The remainder, when strained, might have been five pounds, not more. The market price for it is about ten cents per pound; amount fifty cents. We will say the new hive kept through the winter to receive the bees in the spring contained fifteen pounds; this would also have averaged about ten cents per pound, amounting to $1.50. All that a stock of this kind costs me appears to be just $2.00, and worth at least $5.00. The advantage in changing twenty would be $60.00. The labor of transferring will offset against the trouble of straining, preparing, and the expense of getting the honey to market.

ANOTHER METHOD OF UNITING TWO FAMILIES.

I have occasionally adopted yet another method of making a good stock from two poor ones, which the reader may prefer. When all your old stocks have been reinforced that need it, and you still have some swarms with too few bees and too little honey for safety as they are, two or more can be united. The fact, which has been thoroughly tested, that two families of bees, when united and wintered in one hive, will consume but little, if any more, than each of them would separately, is a very important principle in this matter. If each family should have fifteen pounds of honey, they would consume it all, and probably starve at last, after eating thirty pounds. But if the contents of both were in one hive, it would be amply sufficient, and some to spare in the spring.

UNITING COMB AND HONEY AS WELL AS BEES.

The process of uniting them is simple. Smoke both the stocks or swarms thoroughly, and turn them over. Choose the one with the straightest combs, or the one nearest full, to receive the contents of the other; trim off the points of the combs to make them square across, and this one is ready; remove the sticks from the other, and with your tools take out the combs with the bees on as before directed, one at a time, and carefully set them on the edges of the other; if the shape will admit it, let the edges match; if not, let them cross. Small bits of wood or rolls of paper will be needed between them, to preserve the right distance. When both hives are of one size, the transferred combs will exactly fit, if you are careful to place them as they were before. You will now want to know, "what is to prevent these combs from falling out when the hive is turned over?" This hive is to remain bottom up in some dark place for some time, or till spring. (See method of wintering bees.) The bees will immediately join these combs fast; the hive being inverted, the honey in these combs will be consumed first; and when the hive is again set out in spring, it will be a rare occurrence for any pieces to drop out. Should any pieces project beyond the bottom of the hive, they may be trimmed off even after they are fastened, any time before setting out. An additional cross-stick may pass under the bottom of the combs, to assist in holding them, if you desire. You will probably never discover any difference in the subsequent prosperity in consequence of the joining or crossing of the combs in the middle. I have had them in this way, when they were among the most prosperous of my stocks. As this operation is to be put off till November, it will be an advantage in another way; that is, families of the same apiary can be united, and will mostly forget the old location by spring, and no difficulty arise by returning to the old stand, etc.

WHEN FEEDING SHOULD BE DONE FOR STOCK HIVES.

In some sections of country the *honey* is more frequently wanting than bees, or comb, and some seasons in this; in such cases, it will be found an advantage to feed, until enough is stored for winter. This should be done in September or October. But if they lack comb as well as honey, and you wish to try feeding, (which I seldom do lately,) it should be done if possible in warm weather, as they cannot work combs to advantage in the cold. While feeding bees, it requires a great deal of caution to prevent others from scenting the honey, and their contentions about it. The safest place is on the top of the hive, with a good cap over; but they will not work quite as fast, especially if the weather is cool. The next best place is under the bottom in the manner described in Chapter IX.

Setting out honey to feed all at once, I condemn wholly. These disadvantages attend it: strong stocks that do not need an ounce, will get two or three pounds, while those weaker ones, needing it more, will not get one. Nearly every stock, in a short time, will be fighting. Probably the first bee that comes home with a load, will inform a number of its fellows that a treasure is close at hand. A number will sally out immediately, without waiting for particular directions for finding it; and mistaking other hives for the place, alight there, are seized and probably dispatched. As soon as the honey given them is gone, the tumult is greatly increased, and great numbers are destroyed. If any of your neighbors near you have bees, you must expect to divide with them.

If the honey to be fed is in the comb, and your hives are not full, and they are to be wintered in the house, bottom up, it may be done at any time through the winter, merely by laying pieces with honey on those in the hive. The bees readily remove the contents into their own combs; when empty, remove them and put in more until they have a full supply. They will join such pieces of comb to their own; yet there will be no harm in breaking them loose. The principal objection to feeding in this way, will be found in the tendency to make them uneasy and disposed to leave the hive, when we want them as quiet as possible, A thin muslin cloth, or other means, will be necessary to confine them to the hive.

I have now given directions to avoid killing any family of bees worth saving, if we choose.

When such as need feeding have been fed, and all weak families made strong by additions, etc., but little more fall work is needed in the apiary. It is only when you have weak stocks, unfit to winter, that it is necessary to be on the lookout every warm day to prevent pillage.

CHAPTER XXII.

WINTERING BEES.

There is almost as much diversity of opinion with respect to wintering bees as in the construction of hives, and about as difficult to reconcile.

DIFFERENT METHODS HAVE BEEN ADOPTED.

One will tell you to keep them warm, another to keep them cold; to keep them in the sun, out of the sun, bury them in the ground, put them in the cellar, the chamber, wood-house, and other places, and no places at all; that is, to let them remain as they are, without any attention. Here are plans enough to drive the inexperienced into despair. Yet I have no doubt but that bees have been sometimes successfully wintered by all these contradictory methods. That some of these methods are superior to others, needs no argument to illustrate. But what method *is best*, is our province to inquire. Let us endeavor to examine the subject without prejudice to bias our judgment.

THE IDEA OF BEES NOT FREEZING HAS LED TO ERRORS IN PRACTICE.

By close observation we shall probably discover that the assertion so often repeated, that bees have never frozen except when without honey, has led to an erroneous practice.

APPEARANCE OF BEES IN COLD WEATHER.

We will first endeavor to examine the condition of a stock left to nature, without any care, and see if it affords any hints for our guidance, when to assist and protect with artificial means.

Warmth being the first requisite, a family of bees at the approach of cold weather crowd together in a globular form, into a compass corresponding to the degree of cold; when at zero it is much less than at thirty above. Those on the outside of this cluster are somewhat stiffened with cold; while those inside are as brisk and lively as in summer. In severe weather every possible space within their circle is occupied; even each cell not containing pollen or honey will hold a bee. Suppose this cluster is sufficiently compact for mutual warmth, with the mercury at 40, and a sudden change brings it down to zero, in a few hours, this body of bees, like most other things, speedily contracts by the cold. The bees on the outside, being already chilled, a portion of them that does not keep up with the shrinking mass, is left exposed at a distance from their fellows, and receive but little benefit of the warmth generated there; they part with their vitality, and are lost.

HOW PART OF THE SWARM IS FROZEN.

A good family will form a ball or circle about eight inches in diameter, generally about equal every way, and must occupy the spaces between four or five combs. As combs must separate them into divisions, the two outer ones are smallest, and most exposed of any; these are often found frozen to death in severe weather. Should evidence be wanting from other sources to show that bees will freeze to death, the above would seem to furnish it. It is said, "that in Poland bees are wintered in a semi-torpid state, in consequence of the extreme cold." We must either doubt the correctness of this relation, or suppose the bee of that country a different insect from ours—a kind of semi-wasp, that will live through the winter, and eat little or nothing. The reader can have no difficulty in deciding which is the most probable, whether *bees are bees* throughout the world, endowed with the same faculties and instincts, or that the facts as they are, are not precisely given, especially when we see what our own apiarians tell us about their never freezing.

Here I might use strong language in contradiction; but as I am aware that such a course is not always the most convincing, I prefer the test of close observation. If bees will freeze, it is important to know it, and in what circumstances.

HOW A SMALL FAMILY MAY ALL FREEZE.

Suppose a quart of bees were put in a box or hive where all the cells were filled and lengthened out with honey; the spaces between the combs would be about one-fourth of an inch—only room for one thickness of bees to spread through. The combs would perhaps be one and a half or two inches thick. All the warmth that could be generated then, would be by one course or layer of bees, an inch and a half apart. Although every bee would have food in abundance without changing its position, the first turn of severe weather would probably destroy the whole. This, it may be said, "is an unnatural situation." I will admit that it is; the case was only supposed for illustration. I know that their winter quarters are among the brood combs, where the hatching of the brood leaves most of the cells empty; and the space between the combs is half an inch; a wise and beautiful arrangement; as ten times the number of bees can pack themselves within a circle of six inches, as can in the other case; and in consequence the same number of bees can secure much more animal heat, and endure the cold much better; but a *small* family, even here, will often be found frozen, as well as starving.

FROST AND ICE SOMETIMES SMOTHER BEES.

Besides freezing, there are other facts to be observed in stocks which stand in the cold. If we examine the interior of a hive containing a medium-sized swarm, on the first severely cold morning, except in the immediate vicinity of the bees, we shall find the combs and sides of the hive covered with a white frost. In the middle of the day, or as soon as the temperature is slightly raised, this begins to melt,—first next to the bees, then at the sides. A succession of cold nights will prevent the evaporation of this moisture; and this process of freezing and thawing, at the end of a week or two, will form icicles sometimes as large as a man's finger, attached to the combs and the sides of the hive. When the bottom of the hive is close to the floor, it forms a sealing around the edges, perfectly air-tight, and your bees are smothered. I have frequently heard bee-keepers say in these cases, "The storm blew in, and formed ice all round the bottom, and froze my bees to death." Others that have had their bees in a cold room, finding them thus, "could not see how the water and ice could get there any way; were quite sure it was not there when carried in," &c. Probably they never dreamed of its being accounted for philosophically, and to analyze anything pertaining to bees would be rather small business. But what way can it be accounted for?

FROST AND ICE IN A HIVE ACCOUNTED FOR.

Physiologists tell us "that innumerable pores in the cuticle of the human body are continually throwing off waste or worn out matter; that every exhalation of air carries with it a portion of water from the system, in warm weather unperceived, but will be condensed into particles large enough to be seen in a cold atmosphere." Now, if analogy be allowed here, we will say the bee throws of waste matter and water in the same way. Its food being liquid, nearly all will be exhaled—in moderate weather it will pass off, but in the cold it is condensed—the particles lodge on the combs in form of frost, and accumulate as long as the weather is very severe, a portion melting in the day, and freezing again at night.

THE EFFECT OF ICE OR FROST ON BEES AND COMB.

When the bees are not smothered, this water in the hive is the source of other mischief. The combs are quite certain to mould. The water mould or dampness on the honey renders it thin, and unhealthy for the bees, causing dysentery, or the accumulation of fæces that they are unable to retain. When the hive contains a very large family, or a very small one, there will be less frost on the combs,—the animal heat of the first will drive it off; in the latter there will be but little exhaled.

FROST MAY CAUSE STARVATION.

This frost is frequently the cause of medium or small families starving in cold weather, even when there is plenty of honey in the hive. Suppose all the honey in the immediate vicinity of the cluster of bees is exhausted, and, the combs in every direction from them are covered with frost; if a bee should leave the mass and venture among them for a supply, its fate would be as certain as starvation. And without timely intervention of warmer weather, they *must* perish!

OTHER DIFFICULTIES.

Should they escape starving, there is another difficulty often attending them in continued cold weather. I said that small families exhaled but little. Let us see if we can explain the effect.

There is not sufficient animal heat generated to exhale the aqueous portion of their food. The philosophy that explains why a man in warm blood and in profuse perspiration would throw off or exhale more moisture than in a quiet state, will illustrate this. The bees in these circumstances must retain the water with the excrementitious part, which soon distends their bodies to the utmost, rendering them unable to endure it long. Their cleanly habits, that ordinarily save the combs from being soiled, is not a sure protection now, and they are compelled to leave the mass very often in the severest weather, to expel this unnatural accumulation of fæces. It is frequently discharged even before leaving the comb, but most of it at the entrance; also some scattered on the front side of the hive, and a short distance from it. In a moderately warm day, more bees will issue from a hive in this condition than from others; it appears that a part of them are unable to discharge their burden—their weight prevents their flying—they get down and are lost. When cold weather is too long continued, they cannot wait for warm days to leave, but continue to come out at any time; and not one of such can then return. The cluster inside the hive is thus reduced in numbers till they are unable to generate heat sufficient to keep from freezing. With the indications attendant upon such losses, my own observation has made me somewhat familiar, as the following conversation will illustrate.

FURTHER ILLUSTRATIONS.

A neighbor who wished to purchase some stock hives in the fall, requested my assistance in selecting them. We applied to a perfect stranger; his bees had passed the previous winter in the open air. I found on looking among them that he had lost some of them from this cause, as the excrement was yet about the entrance of one old weather-beaten hive, that was now occupied by a young swarm, and was about half filled with combs.

I saw at once what had been the matter, and felt quite confident that I could give its owner a correct history of it. "Sir," said I, "you have been unfortunate with the bees that were in this hive last winter; I think I can give you some particulars respecting it."

"Ah, what makes you think so? I would like to hear you guess; to encourage you, I will admit that there has been something rather peculiar about it."

"One year ago you considered that a good stock-hive; it was well filled with honey, a good family of bees, and two or three years old or more. You had as much confidence in its wintering as any other; but during the cold weather, somehow, the bees unaccountably disappeared, leaving but a very few, and they were found frozen to death. You discovered it towards spring, on a warm day. When you removed the combs, you probably noticed a great many spots of excrement deposited on them, as well as on the sides of the hive, particularly near the entrance. Also one-half or more of the breeding cells contained dead brood, in a putrid state; and this summer you have used the old hive for a new swarm."

"You are right, sir, in every particular. Now, I would like to know what gave you the idea of my losing the bees in that hive? I can see nothing peculiar about that old hive, more than this one," pointing to another that also contained a new swarm. "You will greatly oblige me if you will point out the signs particularly."

"I will do so with pleasure" (feeling quite willing to give him the impression that I was "posted up" on this subject, notwithstanding it savored strongly of boasting).

I then directed his attention to the entrance in the side of the hive, where the bees had discharged their fæces, on the moment they issued, until it was near the eighth of an inch thick, and two or three inches broad; that yet remained, and just began to cleave off. "You see this brown substance around this hole in the hive?"

"Yes, it is bee-glue (*propolis*); it is very common on old hives."

"I think not; if you will examine it closely, you will perceive it is not so hard and bright; it already begins to crumble; bee-glue is not affected by the weather for years."

"Just so, but what is it, and what has that to do with your guess-work?"

"It is the excrement of the bees. In consequence of a great many cells containing dead brood, which the bees could not enter, they were unable to pack themselves close enough to secure sufficient animal heat to exhale or drive off the water in their food, it was therefore retained in their bodies till they were distended beyond endurance—they were unable to wait for a warm day—necessity compelled them to issue daily during the coldest weather, discharging their fæces the moment of passing the entrance, and part of them before. They were immediately chilled, and could not return; the quantity left about this entrance shows that a great many must have come out. That they came out in cold weather is proved by its being left on the hive, because in warm weather they *leave* the hive for this purpose."

"This is a new idea; at present it seems to be correct; I will think it over. But how did you know that it was not a new swarm; that it was well filled?"

"When looking under it just now, I saw that combs of a dark color had been attached to the sides near the bottom, below where those are at present; this indicates that it had been full, and the dark color that it was not new. Also, a swarm early and large enough to fill such a hive the first season, would not be very likely to be affected by the cold in this way."

"Why not? I think this hive was crowded with bees as much as any of my new swarms."

"I have no doubt they appeared so; but we are very liable to be deceived in such cases, by the dead brood in the combs. A moderate-sized family in such a hive will make more show than some larger ones that have empty cells to creep into, and can pack closer."

"But how did you know about the dead brood?"

"Because old stocks are thus often reduced and lost."

"What were the indications of its being filled with honey?"

"Combs are seldom attached to the side of the hive farther down than they are filled with honey. In this hive the combs had been attached to the bottom, consequently must have been full. Another thing, unless the family is very much reduced, the hive is generally well stored, even when diseased."

"Why did you suppose it was near spring before I discovered it?"

"I took the chances of guessing. The majority of bee-keepers, you know, are rather careless, and when they have fixed their bees for winter, seldom give them much more attention, till they begin to fly out in the spring."

"But what should I have done had I discovered the bees coming out?"

"As it was affected with dead brood, it was but little use to do anything; you would have lost it eventually. But if it had been a stock otherwise healthy, and was affected in this way only because it was a small family, or the severity of the weather, you could have taken it to a warm room, and turned it bottom up; the animal heat would then convert the most of the water contained in their food into vapor; that would rise from the hive, and the bees could retain the excrementitious portion without difficulty till spring."

"I suppose you must get along without losing many through the winter, if I may judge by your confident explanations."

"I can assure you I have but little fear on this head. If I can have the privilege of selecting proper stocks, I will engage not to lose one in a hundred."

"How do you manage? I would be glad to obtain a method in which I could feel as perfectly safe as you appear to."

"The first important requisite is to have all good ones to start with. Enough weak families are united together till they are strong, or some other disposition made of them." I then gave him an outline of my method of wintering, which I can confidently recommend to the reader.

ACCUMULATION OF FÆCES DESCRIBED BY SOME WRITERS AS A DISEASE.

This accumulation of fæces is considered by many writers as a disease—a kind of dysentery. It is described as affecting them towards spring, and several remedies are given. Now if what I have been describing is not the dysentery, why I must think I never had a case of it; but I shall still persist in guessing it to be the same, and suppose that inattention with many must be the reason that it is not discovered in cold weather, at the time that it takes place. Some stocks may be badly affected, yet not lost entirely, when moderate weather will stop its progress. When a remedy is applied in the spring, long after the cause ceases to operate, it would be singular if it was not effectual. I have no doubt but some have taken the natural discharge of fæces, that always takes place in spring when the bees leave the hive, for a disease. Others, when looking for a cause for diseased brood, and found the combs and hive somewhat besmeared, have assigned this as sufficient; but according to my view, have reversed it, giving the effect before the cause.

THE AUTHOR'S REMEDY.

For a time, I supposed that this moisture on the combs gradually mixed with the honey, making it thin, and that the bees eating so much water with their food, would affect them as described. Some experiments that followed, induced me to assign cold as the cause, as I always found, when I put them where it was sufficiently warm, that an immediate cure was the result, or at least, it enabled them to retain their fæces till set out in the spring.

BURYING BEES.

Burying bees in the earth below the frost, has been recommended as a superior method of wintering, for small families. I have known it confidently asserted, that they would lose nothing in weight, and no bees would die. I found, in testing it, that a medium quantity of honey sufficed, and but very few were lost, perhaps less than by any other method. Yet the combs were mouldy, and unfit for further use. There was no escape for the vapor and dampness of the earth. This did not satisfy me; it only cured "one disease by instituting another." I saved the bees, (and perhaps some honey), but the combs were spoiled.

EXPERIMENTS OF THE AUTHOR TO GET RID OF THE FROST.

I wished to keep them warm, and save the bees as well as honey, and at the same time, get rid of the moisture. I found that a large family expelled it much better than small ones; and if all were put together in a close room, the animal heat from a large number combined, would be an advantage to the weak ones, at least,—this proved of some benefit. Yet I found on the sides of a glass hive, that large drops of water would stand for weeks.

SUCCESS IN THIS MATTER.

The following suggestion then came to my relief. If this hive was bottom up, what would prevent all this vapor as it arises from the bees from passing off? (It always rises when warm, if permitted.) The hive was inverted; in a few hours the glass was dry.

This was so perfectly simple, that I wondered I had not thought of it before, and wondered still more that some one of the many intelligent apiarians had never discovered it. I immediately inverted every hive in the room, and kept them in this way till spring; when the combs were perfectly bright, not a particle of mould to be seen, and was well satisfied with the result of my experiment. Although I was fearful that more bees would leave the hives when inverted, than if right side up, yet the result showed no difference. I had now tried both methods, and had some means of judging.

BEES WHEN IN THE HOUSE SHOULD BE KEPT PERFECTLY DARK.

When not kept perfectly dark, a few would leave the hives in either case. I have found it much better to make the room dark to keep the bees in the hive, than to tie over them a thin muslin cloth, as that prevents a free passage of the vapor, and a great number of full stocks were not at all satisfied in confinement; and were continually worrying, and biting at the cloth, till they had made several holes through it for passages out. Thus the little good was attended by an evil, as an offset. Even wire cloth put over to confine them, which would be effectual, would not save bees enough to pay expense. I have thus wintered them for the last ten years, and am extremely doubtful if a better way can be found.[17] For several years I made use of a small bed-room in the house, made perfectly dark, in which I put about 100 stocks. It was lathed and plastered, and no air admitted, except what might come through the floor. It was single, and laid rather close, though not matched.

A ROOM MADE FOR WINTERING BEES.

In the fall of 1849 I built a room for this purpose; the frame was eight by sixteen feet square, and seven high, without any windows. A good coat of plaster was put on the inside, a space of four inches between the siding and lath was filled with saw-dust; under the bottom I constructed a passage for the admission of air, from the north side; another over head for its exit, to be closed and opened at pleasure, in moderate weather, to give them fresh air, but closed when cold, and so arranged as to exclude all the light.

A partition was extended across near the centre. This was to prevent disturbing the whole by letting in light when carrying them out in the spring. By closing the door of this partition, those in one room only need be disturbed at once.

MANNER OF STOWING AWAY BEES.

Shelves to receive the hives were arranged in tiers one above the other; they were loose, to be taken down and put up at pleasure. Suppose we begin at the back end: the first row is turned directly on the floor, a shelf is then put across a few inches above them, and filled, and then another shelf, still above, when we again begin on the floor, and continue thus till the room is full; or if the room is not to be filled, the shelves may be fixed around the sides of the room in two or three courses. This last arrangement will make it very convenient to inspect them at any time through the winter, yet they should be disturbed as little as possible. The manner of stowing each one is to open the holes in the top, then lay down two square sticks, such as are made by splitting a board, of suitable length, into pieces about an inch wide. The hive is inverted on these; it gives a free circulation through the hive, and carries off all the moisture as fast as generated.

TEMPERATURE OF ROOM.

The temperature of such a room will vary according to the number and strength of the stocks put in; 100 or more would be very sure to keep it above the freezing point at all times. Putting a very few into such a room, and depending on the bees to make it warm enough, would be of doubtful utility. If these means will not keep the proper temperature, probably some other method would be better. All full stocks would do well enough, as they would almost any way. Yet I shall recommend housing them whenever practicable. If the number of stocks is few, let the room be proportionably small.[18] It is the smallest families that are most trouble: if they are too cold, it may be known by bees leaving the hive in cold weather, and spots of excrement on the combs; they should then have some additional protection; close part or all of the holes in the top, cover the open bottom partially or wholly, and confine to the hive as much as possible the animal heat; when these means fail, it may be necessary to take them to a warm room, during the coldest weather.

TOO MUCH HONEY MAY SOMETIMES BE STORED.

After the flowers fail, and all the brood has matured and left the combs, it sometimes happens that a stock has an opportunity of plundering, and rapidly filling all those cells that had been occupied with brood during the yield of honey, and which then effectually prevents their storing in them. This, then, prevents close packing, which is all-important for warmth. Although a large family, as much care is needed as with the smaller ones. Also such as are affected with diseased brood should receive extra attention for the same reason.

Some bee-keepers are unwilling to risk the bold measure of inverting the hive, but content themselves by merely opening the holes in the top; this is better than no ventilation, but not so effectual, as all of the moisture cannot escape. There are some who cannot divest themselves of the idea, that if the hive is turned over, the bees must also stand on their heads all winter!

Rats and mice, when they find their way into such room, are less bold with their mischief than if the hive is in its natural position.

MANAGEMENT OF ROOM TOWARDS SPRING.

A few warm days will often occur, towards spring, before we can get our bees out. In these cases, a bushel or two of snow or ice pounded up should be spread on the floor; it will absorb and carry off as it melts much of the heat, that is now unnecessary, and will keep them quiet much longer than without it; (provision for getting rid of this water should be made when putting down the floor.)

TIME FOR SETTING OUT BEES.

The time for carrying out bees is generally in March, but some seasons later. A warm pleasant day is the best, and one quite cold, better than one only *moderately* warm.

After their long confinement, the light attracts them out at once, (unless very cold air prevents), and if the rays of a warm sun do not keep them active, they will soon be chilled and lost.

Some bee-keepers take out their stocks at evening. If we could be always sure of having the next day a fair one, it would probably be the best time; but should it be only moderate, or cloudy, it would be attended with considerable loss—or if the next day should be quite cold, but few would leave, and then the only risk would be to get *a good day*, before one that was just warm enough to make them leave the hive, but not quite enough to enable them to return.

NOT TOO MANY STOCKS TAKEN OUT AT ONCE.

When too many are taken out at once, the rush from all the hives is so much like a swarm, that it appears to confuse them. Some of the stocks by this means will get more bees than actually belong to them, while others are proportionably short, which is unprofitable, and to equalize them is some trouble; yet it may be done. Being all wintered in one room, the scent or the means of distinguishing their own family from strangers, becomes so much alike, that they mix together without contention.

FAMILIES MAY BE EQUALIZED.

By taking advantage of this immediately, or before the scent has again changed, and each hive has something peculiar to *itself*, you can change the stands of very weak and very strong families.

To prevent, as far as possible, some of these bad effects, I prefer waiting for a fair day to begin, and then not until the day has become sufficiently warm to make it safe from chill.

SNOW NEED NOT ALWAYS PREVENT CARRYING OUT BEES.

I am not particular about the snow being gone—if it has only lain long enough to have melted a part of it, it is "terra firma" to a bee, and answers equally well as the bare earth. When the day is right, about ten o'clock I put out twelve or fifteen, taking care that each hive occupies its old stand, at the same time endeavoring to take such as will be as far apart as possible; (to make this convenient, they should be carried in in the manner that you wish them to come out.) When the rush from these hives is over, and the majority of the bees has gone back, I set out as many more about twelve o'clock, and when the day continues fair, another lot about two. In the morning, while cool, I move from the back to the first apartment, about as many as I wish to set out in a day, except a few at the last.

To do this in the middle of the day, while warm, would induce a good many bees to leave the hive, while the light was admitted, and which would be lost. It will be supposed generally that their long confinement makes them thus impatient to get out; but I have frequently returned stocks during a cold turn of weather after they had been out, and always found such equally as anxious to come out, as those which had been confined throughout the winter; without the airings, I have kept them thus confined, for five months, without difficulty! The important requisites are, sufficient warmth and perfect darkness.

DOES NOT ANALOGY PROVE THAT BEES SHOULD BE KEPT WARM IN WINTER?

Opposition to this method of wintering will arise with those who have always thought that bees must be kept cold; "the colder the better." I would suggest for their consideration the possibility of some analogy between bees and some of the warm-blooded animals—the horse, ox, and sheep, for instance, that require a constant supply of food, that they may generate as much caloric as is thrown off on the cold air. This seems to be regulated by the degree of cold, else why do they refuse the large quantity of tempting provender in the warm days of spring, and greedily devour it in the pelting storm? The fact is pretty well demonstrated, that the quantity of food needed for the same condition in spring, is much less when protected from the inclemency of the weather, than when exposed to the severe cold. The bee, unlike the wasp, when once penetrated with frost, is dead—*their temperature must be kept considerably above the freezing point, and to do this, food is required.* Now if the bees are governed by the same laws, and cold air carries off more heat than warm, and their source of renewing it is in the consumption of honey in proportion to the degree of cold, common sense would say, keep them warm as possible. As a certain degree of heat is necessary in all stocks, it may take about such a quantity of honey to produce it, and this may explain why a small family requires about the same amount of food as others that are very large.

THE NEXT BEST PLACE FOR WINTERING BEES.

A *dry*, warm cellar is the next best place for wintering them; the apiarian having one perfectly dark, with room to spare, will find it a very good place, in the absence of a room above ground. If a large number was put in, some means of ventilation should be contrived for warm turns of weather. I know an apiarian, who by my suggestion has wintered from sixty to eighty stocks in this way, for the last six years, with perfect success, not having lost one. Another has wintered thirty with equal safety.

As for burying them in the earth, I have not the least doubt, if a dry place should be selected, the hive inverted, and surrounded with hay, straw, or some substance to absorb the moisture, and protected from the rain, at the top of the covering, that perfect success would attend the experiment. But this is only theory; when I tried the experiment of burying, and had the combs mould, the hives ware right side up.

EVILS OF WINTERINGS IN THE OPEN AIR CONSIDERED.

As a great many bee-keepers will find it inconvenient, or be unable to avail themselves of my method of wintering, it will be well enough to see how far the evils of the open air, which we have already glanced at, may be successfully avoided. I am told by those who have tried wintering them in straw hives, that in this respect they are much safer than those made of boards; probably the straw will absorb the moisture. But as these hives are more trouble to construct, and their shape will prevent the use of suitable boxes for surplus honey, this one advantage will hardly balance the loss. They are said also to be more liable to injury from the moth. We want a hive that will unite advantageously as many points as possible.

It should be remembered that bees always need air, especially in the cold.[19] With this in view, we will try to dispose of the vapor or frost. If the hive is raised sufficient to let it out, it will let in the mice; to prevent which, it should be raised only about one-fourth of an inch. The hole in the side should be nearly covered with wire cloth to keep out the mice; but give a passage for the bees; otherwise they collect here, endeavoring to get out, and remain till chilled, and thus perish by hundreds. The boxes on the top must be removed, but not the cap or cover; the holes all opened, to let the vapor pass up into the chamber; if this is made with perfectly close joints, so that no air escapes, it should be raised a very little; otherwise not. The moisture will condense on the sides and top, when it melts will follow the sides to the bottom, and pass out; the rabbeting around the top of the hive will prevent its getting to the holes, and down among the bees. It will be easily comprehended, that a hole between each two combs at the top, (as mentioned in the subject of putting on the boxes,) will ventilate the hive much better than where there is but one or two, or where there is a row of several, and all are between two combs.

BUT LITTLE RISK WITH GOOD STOCKS.

All *good stocks* may be wintered in this way, with but little risk in most situations. Whether in the bleak north-wind, buried in a snow-bank, or situated warm and pleasant, it will make no great difference. The mice cannot enter; the holes give them air, and carry off moisture, &c. But second-rate stocks are not equally safe in cold situations.

EFFECT OF KEEPING SECOND-RATE STOCKS OUT OF THE SUN.

It has been strongly urged, without regard to the strength of the stock, to keep them all out of the sun; because an occasional warm day would call out the bees, when they get on the snow, and perish; this is a loss, to be sure, but there is such a thing as inducing a greater one by endeavoring to avoid this. I have said in another place that second rate or poor stocks might occasionally starve, with plenty of stores in the hive, on account of frosty combs. If the hive is kept from the sun, in the cold, the periods of temperate weather might not occur as often, as the bees would exhaust the honey within their circle or cluster. But on the contrary, when the sun can strike the hive, it warms up the bees, and melts the frost more frequently. The bees may then go among their stores and obtain a supply, generally, as often as needed. We seldom have a winter without sunny days enough for this purpose; but should such an one occur, stocks of this class should be brought into a warm room, once in four or five days, for a few hours at a time, to give them a chance to get at the honey. Stocks much below second-rate cannot be wintered successfully in this climate; the only place for them is the warm room. I have known bees thoroughly covered in a snow-drift, and their owner was at considerable trouble to shovel the snow away, fearing it would smother them. This is unnecessary, when protected from the mice and ventilated as just directed; a snow-bank is about as comfortable a place as they can have, except in the house. When examined a short time after being so covered, the snow for a space of about four inches on every side of the hive is found melted, and none but quite poor stocks would be likely to suffer with this protection. A little snow around the bottom, without a vent in the side of the hive, might smother them.

EFFECTS OF SNOW CONSIDERED.

As for bees getting on the snow, I apprehend that not many more are lost there, than on the frozen earth; that is, in the same kind of weather. I have seen them chilled, and lost on the ground by hundreds, when a casual observer would not have noticed them; whereas, had they been on the snow, at the distance of several rods, every bee would have been conspicuous. Snow is not to be dreaded as much as chilly air. Suppose a hive stands in the sun throughout the winter, and bees are allowed to leave when they choose, and a portion are lost on the snow, and that it was possible to number all that were lost by getting chilled, throughout the season, on the bare earth— the proportion (in my opinion) lost on the snow would not be one in twenty. A person that has not closely observed during damp or chilly weather, in April, May, or even the summer months, has no adequate conception of the number. Yet, I do not wish to be understood that it is of no consequence what are lost on the snow, by any means. On the contrary, a great many are lost, that might be saved with proper care. But I would like to impress the fact, that frozen earth is not safe without warm air, any more than snow, when crusted, or a little hard. Even when snow is melting, it is solid footing for a bee; they can and do rise from it, with the same ease as from the earth. Bees that perish on snow in these circumstances, would be likely to be lost if there was none.

STOCKS TO BE PROTECTED ON SOME OCCASIONS.

The worst time for them to leave the hive is immediately after a new snow has fallen, because if they light on it then, it does not sustain their weight; and they soon work themselves down out of the rays of the sun, and perish. Should it clear off pleasant, after a storm of this kind, a little attention will probably be remunerated. Also, when the weather is moderately warm, and not sufficiently so to be safe, they should be kept in, whether snow is on the ground or otherwise.

For this purpose, a wide board should be set up before the hive to protect it from the sun, at least above the entrance in the side. But if it grows sufficiently warm so that bees leave the hive when so shaded, it is a fair test by which to tell when it will do to let them have a good chance to sally out freely, except in cases of a new snow, when it is advisable to confine them to the hive. The hive might be let down on the floor-board, and the wire-cloth cover the passage in the side, and made dark for the present; raising the hive at night again, as before. I have known hundreds of stocks wintered successfully without any such care being taken, and the bees allowed to come out whenever they chose to do so. Their subsequent health and prosperity proving that it is not altogether ruinous. It has been recommended to enclose the whole hive by a large box set over it, and made perfectly dark, with means for ventilation, &c. (A snow-bank would answer equally well, if not better.) For large families it would do well enough, as would also other methods. But I would much rather take the chances of letting them all stand in the sun, and issue as they please, than to have the warmth of the sun entirely excluded from the moderate-sized families. I never knew a whole stock lost by this cause alone.[20] Yet, I have known a great many starved, merely because the sun was not allowed to melt the frost on the combs, and give them a chance to get at their stores.

DO THE BEES EAT MORE WHEN ALLOWED TO COME OUT OCCASIONALLY IN WINTER?

Besides the loss of bees on the snow when standing in the sun, and taking an airing occasionally, there are some economical bee-keepers who urge this disadvantage, "that every time bees come out in winter they discharge their excrement, and eat more honey in consequence of the vacant room." What a ridiculous absurdity it would be to apply this principle to the horse, whose health, strength, and vital heat is sustained by the assimilation of food! and the farmer is not to be found who would think of saving his provender by the same means. That bees are supported in cold weather on the same principle is indicated strongly, if not conclusively.

Is it not better (if what has been said on the subject of wintering bees is correct) to keep our bees warm and comfortable when practicable, as a means of saving honey?

To winter bees in the best manner, considerable care is required. Whenever you are disposed to neglect them, you should bear in mind that one early swarm is worth two late ones; their condition in spring will often decide this point. Like a team of cattle or horses when well wintered, they are ready for a good season's work, but when poorly wintered have to recruit a long time before they are worth much.

CHAPTER XXIII.

SAGACITY OF BEES.
ARE NOT BEES DIRECTED ALONE BY INSTINCT?

On this subject I have but little to say, as I have failed to discover anything uncommonly remarkable, separate and distinct in one swarm, that another would not exhibit. I have found one swarm guided alone by instinct, doing just what another would under the same circumstances.

Writers, not contented with the astonishing results of instinct, with their love of the marvellous, must add a good share of reason to their other faculties,— "an adaptation of means to ends, that reason alone could produce." It is very true, without close inspection, and comparing the results of different swarms in similar cases, one might arrive at such conclusion. It is difficult, as all will admit, "to tell where instinct ends, and reason begins." Instances of sagacity, like the following, have been mentioned. "When the weather is warm, and the heat inside is somewhat oppressive, a number of bees may be seen stationed around the entrance, vibrating their wings. Those inside will turn their heads towards the passage, while those outside will turn theirs the other way. A constant agitation of air is thus created, thereby ventilating the hive more effectually." *All full stocks do this in hot weather.*

WHAT THEY DO WITH PROPOLIS.

"A snail had entered the hive and fixed itself against the glass side. The bees, unable to penetrate it with their stings, the cunning economists fixed it immovably, by cementing merely the edge of the orifice of the shell to the glass with resin, (propolis), and thus it became a prisoner for life." Now the instinct that prompts the gathering of propolis in August, and filling every crack, flaw, or inequality about the hive, would cement the edges of the snail-shell to the glass, and a small stone, block of wood, chip, or any substance that they are unable to remove, would be fastened with it in the same manner. The edges or bottom of the hive, when in close proximity to the bottom, is joined to it with this substance. Whatever the obstacle may be, it is pretty sure to receive a coating of this. The stoppers for the holes at the top are held in their places on the same principle; and the unaccountable sagacity that once fastened a little door, might possibly be nothing more than the same instinct.

Another principle, I think, will be found to be universal with them, instead of sagacious reasoning.

Whenever the combs in a hive have been broken, or when combs have been added, as was mentioned in the chapter on fall management, the first duty of the bees appears to be to fasten them as they are; when the edges are near the side of the hive, or two combs in contact, a portion of wax is detached and used for joining them together, or to the side.

MENDING BROKEN COMBS.

Where two combs do not touch, and yet are close together, a small bar is constructed from one to the other, preventing any nearer approach. (This may be illustrated by turning the hive a few inches from the perpendicular after being filled with combs in warm weather.)

MAKING PASSAGES TO EVERY PART OF THEIR COMBS.

Should nearly all the combs in the hive become detached from any cause, and lie on the bottom in one "grand smash of ruin," their first steps are, as just described, pillars from one to the other to keep them as they are. In a few days, in warm weather, they will have made passages by biting away combs where they are in contact, throughout every part of the mass; little columns of wax below, supporting the combs above,—irregular, to be sure, but as well as circumstances admit. Not a single piece can be removed without breaking it from the others, and the whole will be firmly cemented together. A piece of comb filled with honey, and sealed up, may be put in a glass box with the ends of these cells so sealed, touching the glass. The principle of allowing no part of their tenement to be in a situation inaccessible, is soon manifested. They immediately bite off the ends of the cells, remove the honey that is in the way, and make a passage next to the glass, leaving a few bars from it to the comb, to steady and keep it in its position. A single sheet of comb lying flat on the bottom-board of a populous swarm is cut away under side, for a passage in every direction, numerous little pillars of wax being left for its support. How any person in the habit of watching their proceedings, with any degree of attention, could come at the conclusion that the bees raised such comb by mechanical means and then put under the props for its support, is somewhat singular. Their efforts united for such a purpose like reasonable beings, I never witnessed.

These things, considered as the effect of instinct, are none the less wonderful on that account. I am not sure but the display of wisdom is even greater than if the power of planning their own operations had been given them.

I have mentioned these, to show that a course of action called forth by the peculiar situation of one family, would be copied by another in a similar emergency, without being aware of its ever being done before. Were I engaged in a work of fiction, I might let fancy reign and endeavor to amuse, but this is not the object. Let us endeavor then to be content with truth, and not murmur with its reality. When we take a survey of the astonishing regularity with which they construct their combs without a teacher, and remember that the waxen material is formed in the rings of their body, that for the first time in life, without an experienced leader's direction, they apply a claw to detach it, that they go forth to the fields and gather stores unbidden by a tyrant's mandate, and throughout the whole cycle of their operations, one law and power governs. Whoever would seek mind as the directing power, must look beyond the sensorium of the bee for the source of all we behold in them!

CHAPTER XXIV.

STRAINING HONEY AND WAX.

When about to remove the contents of a hive, I have never found it necessary to use all the precautions often recommended to prevent the access of bees. I have seen it stated that a room in which there was a chimney open, would be unsuitable, as the bees would scent the honey, and thus find their way down into the room. I never was thus troubled by their perpendicular travelling. It is true, if the day was warm, and a door or window was standing open, the bees would find their way in during a scarcity of honey. But with doors and windows closed no difficulty need be apprehended.

METHODS OF REMOVING COMBS FROM THE HIVE.

The most convenient way to remove combs from the hive is to take off one of its sides, but this is apt to split the boards, if it was properly nailed, and injure it for subsequent use. With tools such as have been described, it may be done very nicely, and leave the hive whole. The chisel should have the bevel all on one side, like those used by carpenters. When you commence, turn the flat side next the board of the hive, and the bevel crowded by the combs will follow it close the whole length; with the other tool they are cut across the top, and readily lifted out. If preferred, they may be cut across near the centre and take out half a sheet at a time; this is sometimes necessary on account of the cross-sticks.

DIFFERENT METHODS OF STRAINING HONEY.

Such combs as are taken from the middle or vicinity of brood-cells, are generally unfit for the table; such should be strained. There are several methods of doing it. One is, to mash the comb and put it in a bag, and hang it over some vessel to catch the honey as it drains out. This will do very well for small quantities in warm weather, or in the fall before there is any of it candied. Another method is to put such combs into a colander, and set this over a pan, and introduce it into an oven after the bread is out. This melts the combs. The honey and a portion of the wax run out together. The wax rises to the top and cools in a cake. It is somewhat liable to burn, and requires some care. Many prefer this method, as there is less taste of bee-bread, no cells containing it being disturbed, but all the honey is not certain to drain out without stirring it. If disposed, two qualities may be made, by keeping the first separate. Another method is merely to break the combs finely, and put them into a colander, and allow the honey to drain out without much heat, and afterwards skim off the small particles that rise to the top, or when very particular, pass the honey through a cloth, or piece of lace. But for large quantities, a more expeditious mode is to have a can and strainer, made for the purpose, where fifty pounds or more can be worked out at once. The can is made of tin, twelve or fourteen inches deep, by about ten or twelve diameter, with handles on each side at the top, for lifting it. The strainer is just enough smaller to go down inside the can; the height may be considerably less, providing there are handles on each side to pass out at the top; the bottom is perforated with holes like a colander, combs are put into this, and the whole set into a kettle of boiling water, and heated without any risk of burning, until all the wax is melted, (which may be ascertained by stirring it,) when it may be taken out. All the wax, bee-bread, &c., will rise in a few minutes. The strainer can now be raised out of the top and set on a frame for the purpose, or by merely tipping it slightly on one side it will rest on the top of the can. It might be left to cool before raising the strainer, were it not liable to stick to the sides of the can; the honey would be full as pure, and separate nearly as clean from the wax and bee-bread, &c. When raised out before cooling, the contents should be repeatedly stirred, or considerable honey will remain. Two qualities may be made by keeping the first that runs through separate from the last, (as stirring it works out the bee-bread). Even a third quality maybe obtained by adding a little water, and repeating the process. This is worth but little. By boiling out the water, without burning, and removing the scum, it will do to feed bees. By adding water until it will just bear a potato, boiling and skimming, and letting it ferment, it will make metheglin, or by letting the fermentation proceed it will make vinegar. Honey that has been heated thoroughly, will not candy as readily as when strained without heat. A little water may be added to prevent its getting too hard; but should it get so in cold weather, it can at any time be warmed, and water added until it is of the right consistence.

GETTING OUT WAX—DIFFERENT METHODS.

Several methods have been adopted for separating the wax. I never found any means of getting out the *whole*. Yet I suppose I came as near it as any one. Some recommend heating it in an oven, similar to the method of straining honey through the colander, but I have found it to waste more than when melted with water. A better way for small quantities, is to half fill a coarse stout bag with refuse comb and a few cobble-stones to sink it, and boil it in a kettle of water, pressing and turning it frequently till the wax ceases to rise. When the contents of the bag are emptied, by squeezing a handful, the particles of wax may be seen, and you may thereby judge of the quantity thrown away. For large quantities the foregoing process is rather tedious. It can be facilitated by having two levers four or five feet long and about four inches wide, and fastened at the lower end by a strong hinge. The combs are put into a kettle of boiling water, and will melt almost immediately; it is then put into the bag, and taken between the levers in a wash-tub or other large vessel and pressed, the contents of the bag shaken, and turned, several times during the process, and if need be returned to the boiling water and squeezed again. The wax, with a little water, is now to be remelted and strained again through finer cloth, into vessels that will mould it into the desired shape. As the sediment settles to the bottom of the wax when melted, a portion may be dipped off nearly pure without straining.

Wax in cool weather may be whitened in a short time in the sun, but it must be in very thin flakes; it is readily obtained in this shape by having a very thin board or shingle, which should be first thoroughly wet, and then dipped into pure melted wax; enough will adhere to make it the desired thickness, and will cool instantly on being withdrawn. Draw a knife along the edges, and it will readily cleave off. Exposed to the sun in a window or on the snow, it will become perfectly white, when it can be made into cakes for market, where it commands a much higher price than the yellow. It is said there is a chemical process that whitens it readily, but I am not acquainted with it.

CHAPTER XXV.

PURCHASING STOCKS AND TRANSPORTING BEES.

If the reader has no bees, and yet has had interest or patience to follow me thus far, it is presumptive evidence that he would possess the requisite perseverance to take charge of them. It would be well, however, to remember the anxieties, perplexities, and time necessary to take the proper care, as well as the advantages and profit.

But if you are disposed to try the experiment, very likely some directions for a commencement would be acceptable.

WHY THE WORD LUCK IS APPLIED TO BEES.

There has been so much uncertainty in stock of this kind, that the word *luck* has been made to express too much. Some have been successful, while others have failed entirely; this has suggested the idea that *luck* depended on the manner that the stocks were obtained; and here again there seems to be a variety of opinions, as is the case always, when a thing is guessed at. One will assert that the "fickle dame" is charmed into favor by stealing a stock or two to begin with, and returning them after a start. Another, (a little more conscientious, perhaps) that you must take them without *liberty*, to be sure, but leave an equivalent in money on the stand. Another, that the only way to get up an effectual charm, is to exchange sheep for them; and still another says, that *bees must always be a gift*. I have had all these methods offered me gratis, with gravity, suitable to make an impression. And, finally, there has yet another method been found out, and that is, when you want a few stocks of bees go and buy them, yes, and pay for them too, in dollars and cents, or take them for a share of the increase for a time, if it suits your pecuniary resources best. And you need not depend on any *charm* or mystic power for your success—if you do, I cannot avoid the unfavorable prediction of a failure. It is true that a few have accidentally prospered for a few years; I say accidentally, because when they have no true principles of management, it must be the result of accident. It is a saying with some, that "one man can't have luck but few years at once," and others none at all, although he tries the whole routine of charms. Nearly twenty years ago, when my respected neighbor predicted a "turn in my luck, because it was always so," I could not understand the force of this reasoning, unless it belonged to the nature of bees to deteriorate, and consequently run out. I at once determined to ascertain this point. I could understand how a farmer would often fail to raise a crop, if he depended on chance or luck for success, instead of fixed natural principles. It was possible that bees might be similar. I found that in good seasons the majority of people had luck, but in poor seasons, the reverse, and when two or three occurred in succession, then was the time to lose their luck. It was evident, then, if I could pass in safety the poor seasons by any means, I should do well enough in good ones.[21] The result has given me but little reason to complain. My advice therefore is, that reliance should be placed on proper management, instead of luck, arising from the manner the first stock was obtained. Should any one feel disposed to make you a present of a stock or two of bees, I would advise you to accept the offer and be thankful, discarding all apprehension of a failure on that account. Or if any one is willing you should take some on shares, this is a cheap way to get a start, and you have no risk of loss in the old stock. Yet if bees prosper, the interest on the money that stocks cost is a mere trifle in comparison to the value of increase, and you have the same trouble. On the other hand, the owner of bees can afford to take care of a few hives more, for half the profits, which he has to give if another takes them; this is apt to be the case, especially, with such as have no faith in charms.

RULE IN TAKING BEES FOR A SHARE.

The rule generally adopted for taking bees is this. One or more stocks are taken for a term of years, the person taking them finding hives, boxes, and bestowing whatsoever care is necessary, and returning the old stocks to the owner with half the increase and profits.

A MAN MAY SELL HIS "LUCK."

There are yet a few persons who refuse to sell a stock of bees, because it is "bad luck." There is often some grounds for this notion. It might arise under the following circumstances. Suppose a person has a half dozen hives, three extra good, the others of the opposite extreme. He sells for the sake of the better price his three best; there is but little doubt but his best "luck" would go too! But should his poorest be taken, the result would be different, without doubt.

But there are cases where an apiarian has more stocks than he wishes to keep. (It has been the case with myself frequently.) Persons wishing to sell, are the proper ones of which to buy. Purchasers seldom want any but first-rate stocks, they are generally cheapest in the end. There is usually a difference of about a dollar in the spring and fall prices, and five and six dollars are common charges. I have known them sell at auction at eight, but in some sections they are less.

FIRST-RATE STOCKS RECOMMENDED TO BEGIN WITH.

For a beginning then, I would recommend purchasing none but first-rate stocks; it will make but little difference in the risk, whether you obtain them in the spring, or fall, if you have read my remarks on winter management with attention; I have already said the requisites for a good stock for winter, were a numerous family and plenty of honey, and that the cluster of bees should extend through nearly all the combs, &c. To avoid as far as possible diseased brood, find an apiary where it has never made its appearance, to make purchases. There are some who have lost bees by it, and yet are totally ignorant of the cause. It would be well, therefore, to inquire if any stocks have been lost, and then for the cause—be careful that secondary are not mistaken for primary causes.

OLD STOCKS ARE GOOD AS ANY, IF HEALTHY.

When it appears that all are exempt, (by a thorough examination, if not satisfied without,) you need not object to stocks two or three years old; they are just as good as any, sometimes better, (providing they have swarmed the season previous, according to one author; because such always have young queens, which are more prolific than old ones, that will be in all first swarms).

Old stocks are as prosperous as any, as long as they are healthy, yet they are more liable to become diseased.

CAUTION RESPECTING DISEASED BROOD.

When no apiary from which to purchase can be found, but where the disease *has made* its appearance, and you are necessitated to purchase from such, or not at all, you cannot be too cautious about it. It would be safest in this case to take none but young swarms, as it is not so common for them to be affected the first season, yet they are not always exempt. But here, again, you may not be allowed to take all young stocks; in which case let the weather be pretty cold, the bees will be further up among the combs, and give a chance to inspect the combs. At this season, say not earlier than November, all the healthy brood will be hatched. Sometimes, a few young bees may be left that have their mature shape, and probably had been chilled by sudden cold weather—these are not the result of disease, the bees will remove them the next season, and no bad results follow. In warm weather a satisfactory inspection can be had no other way, but by the use of tobacco smoke. Be particular to reject all that are affected with the disease in the least; better do without, than take such to begin with. (A full description has been given of this disease in another place.)

RESULT OF IGNORANCE IN PURCHASING.

A neighbor purchased thirteen stock-hives; six were old ones, the others swarms of the last season. As the old hives were heavy, he of course thought them good; either he knew nothing of the disease, or took no trouble to examine; five of the six old ones were badly affected. Four were lost outright, except the honey; the fifth lasted through the winter, and then had to be transferred. He had flattered himself that they were obtained very cheaply, but when he made out what his good ones cost, he found no great reason, in this respect, for congratulation.

SIZES OF HIVES IMPORTANT.

Another point is worthy of consideration: endeavor to get hives as near the right size as possible, *viz.*, 2,000 cubic inches; better too large than too small. If too large, they may be cut off, leaving them the proper size. But yet, it often makes an ungainly shape, being too large square for the height. As the shape probably makes no difference in the prosperity of the bees, the appearance is the principal objection, after being cut off.

An acquaintance had purchased a lot of bees in very large hives, and called on me to know what to do with them, as he feared such would not swarm well in consequence; I told him it would be doubtful, unless he cut them off to the right size.

"Cut 'em off! how can that be done? there is bees in 'em."

"So I expected, but it can be done nearly as well as if empty."

"But don't you get stung dreadfully?"

"Not often: if it is to be done in warm weather, I smoke them well before I begin; *in very cold weather* is the best time, then it is unnecessary; simply turn the hive bottom up, mark off the proper size, and with a sharp saw take it off without trouble."

"Some are filled with combs; you don't cut off such, do you?"

"Certainly; I consider all the room for combs in a hive over 2,000 inches as worse than lost."

"What will you ask to cut mine off? If I could see it done once, I might do it next time."

"The charge will be light; but if you intend to keep bees, you should learn to do everything pertaining to them, and not be dependent on any one; I did it before I ever saw or heard of its being done." I then gave him full directions how to manage, but could not persuade him to undertake.

HOW LARGE HIVES CAN BE MADE SMALLER.

A short time after, I attended, on a cold day, with a sharp saw, square, &c. I found his hives fourteen inches square inside, and eighteen deep, holding about 3,500 inches. Of this square, a little more than ten inches in height, would make just the right size. To work convenient, I inverted the hive on a barrel, set on end, marked the length, and sawed it off, without a bee leaving. It was very cold, (mercury at 6 deg.) The bees came to the edges of the combs, but the cold drove them back. In a short time I had taken off six;

four when done were just about full; the other two were so when I began, but they were marked and sawed like the rest; when the combs were attached, they were severed with a knife, and the piece of the hive thus loose, was raised off, leaving several inches of the combs projecting out of the hive. I now cut off the first comb, even with the bottom of the hive. On the next comb there were a few bees; with a quill these were brushed down into the hive; this piece was then removed, and the bees on the other side of it were brushed down also. In this way all others were removed, and left the hive just full. The other full hive, after it was sawed on each side, a small wire was drawn through, parallel with the sheets, and severed all the combs at once; each piece was taken out, and the bees that were clustered on them brushed back; removing the loose part of the hive, was the last thing to be done. This last method was preferred to the other by my employer; yet it was all performed to his satisfaction, no sting or other difficulty about it, except the trouble of warming fingers rather frequently. Tobacco smoke would have kept them quiet during the operation, nearly as well. If preferred, a hive may stand right side up while sawing it.

MODERATE WEATHER BEST TO REMOVE BEES.

In transporting your bees, avoid if possible the two extremes of very cold, or very warm weather. In the latter the combs are so nearly melted, that the weight of the honey will bend them, bursting the cells, spilling the honey, and besmearing the bees. In very cold weather, the combs are brittle, and easily detached from the sides of the hive. When necessitated to move them in very cold weather, they should be put up an hour or so before starting. The agitation of the bees after being disturbed will create considerable heat; a portion of this will be imparted to the combs, and add to their strength.

PREPARATIONS FOR TRANSPORTING BEES.

To prepare for moving them, pieces of thin muslin about half a yard square is as good as anything, secured by carpet tacks.

SECURING BEES IN THE HIVE.

The hive is inverted, and the cloth put over, neatly folded, and fastened with a tack at the corners, and another in the middle. The tack is crowed in about two-thirds of its length, it then presents the head convenient to pull out. If the bees are to go a great distance, and require to be shut up several days, the muslin will be hardly sufficient, as they would probably bite their way out. Something more substantial would then be required. Take a board the size of the bottom, cut out a place in the middle, and cover with wire cloth, (like the one recommended for hiving,) and fasten it with tacks. This board is to be nailed on the hive. After the nails are driven, with the hammer start it off about the eighth of an inch; it will admit a little air around the sides as well as the middle, quite necessary for heavy stocks. But very small families might be safe without the wire cloth; air enough would pass between the hive and board, except in warm weather. New combs break easier than old.

BEST CONVEYANCE.

Probably the best conveyance is a wagon with elliptic springs. But a wagon without springs is bad, especially for young stocks. Yet I have known them moved safely in this way, but it required some care in packing with hay, or straw, under and around them, and careful driving. Good sleighing will answer very well, and by some thought to be the best time.

HIVE TO BE INVERTED.

Whatever conveyance is employed, the hive should be inverted. The combs will then all rest closely on the top, and are less liable to break than when right end up, because then the whole weight of the combs must depend upon the fastenings at the top and sides for support, and are easily detached and fall. When moving bees, so reversed, they will creep upward; in stocks part full, they will often nearly all leave the combs, and get upon the covering. In a short time after being set up, they will return, except in very cold weather, when a few will sometimes freeze; consequently a warm room is required to put them in for a short time.

After carrying them a few miles, the disposition to sting is generally gone, yet there are a few exceptions. In moderate weather, when bees are confined, they manifest a persevering determination to find their way out, particularly after being moved, and somewhat disturbed. I have known them to bite holes through muslin in three days. The same difficulty is often attendant on attempting to confine them to the hive by muslin when in the house in the winter, except when kept in a cold situation. Should any combs become broken, or detached from their fastenings, in hives not full, by moving or other accident, rendering them liable to fall when set up, the hive may remain inverted on the stand till warm weather, if necessary, and the bees have again fastened them, which they do soon after commencing work in the spring. If they are so badly broken that they bend over, rolls of paper may be put between them to preserve the proper distance till secured. When they commence making new combs, or before, it is time to turn the right end up. While the hive is inverted, it is essential that a hole is in the side, through which the bees may work. A board should fit close over the bottom, and covered, to effectually prevent any water from getting among the bees, &c.

CONCLUSION.

In conclusion I would say, that the apiarian who has followed me attentively, and has added nothing of value to his stock of information, possesses an enviable experience that all should strive to obtain.

It has been said that "three out of five who commence an apiary must fail;" but let us suppose it is through ignorance or inattention, and not inherent with the bees. To the beginner then I would say,—if you expect to succeed in obtaining one of the most delectable of sweets for your own consumption, or the profit in dollars and cents, you will find something more requisite than merely holding the dish to obtain the porridge. "SEE YOUR BEES OFTEN," and know at all times their actual condition. This one recipe is worth more than all others that can be given; it is at the head of the class of duties; *all others begin here*. Even the grand secret of successfully combating the worms,—KEEP YOUR BEES STRONG, must take its rise at this point. With the above motto acted upon, carried out fully, and with perseverance, you cannot well fail to realize all reasonable expectations. Avoid over-anxiety for a rapid increase in stocks; try and be satisfied with one good swarm from a stock annually, your chances are better than with more; do not anticipate the golden harvest too soon. You will probably be necessitated to discard some of the *extravagant* reports of profits from the apiary. Yet you will find one stock trebling, perhaps quadrupling its price or value in products, while the one beside it does nothing. In some seasons particularly favorable your stocks collectively will yield a return of one or two hundred per cent.—in others, hardly make a return for trouble. The proper estimate can be made only after a number of years, when, if they have been judiciously managed, and your ideas have not been too extravagant, you will be fully satisfied. I have known a single stock in one season to produce more than twenty dollars in swarms and honey, and ninety stocks to produce over nine hundred dollars, when a few of the number added not a farthing to the amount. I do not wish to hold out inducements for any one to commence bee-keeping, and end it in disgust and disappointment. But I would encourage all suitable persons to try their skill in bee management. I say suitable persons, because there are many, very many, not qualified for the charge. The careless, inattentive man, who leaves his bees unnoticed from October till May, will be likely to complain of ill success.

Whoever cannot find time to give his bees the needed care, but can spend an hour each day obtaining gossip at the neighborhood tavern, is unfit for this business. But he who has a home, and finds his affections beginning to be divided between that and his companions of the bar-room, and wishes to withdraw his interest from unprofitable associates, and yet has nothing of sufficient power to break the bond, to what can he apply with a better prospect of success, than to engage in keeping bees? They make ample returns for each little care. Pecuniary advantages are not all that may be gained—a great many points concerning their natural history are yet in the dark, and many are disputed. Would it not be a source of satisfaction to be able to contribute a few more facts to this interesting subject, adding to the science, and holding a share in the general fund? Supposing all the mysteries pertaining to their economy discovered and elucidated, precluding all chance of further additions, would the study be dry and monotonous? On the contrary, the verification witnessed by ourselves would be so fascinating and instructive, that we cannot avoid pitying the condition of that man who finds gratification only in the gross and sensual. It has been remarked, that "he who cannot find in this and other branches of natural history a salutary exercise for his mental faculties, inducing a habit of observation and reflection, a pleasure so easily obtained, unalloyed by any debasing mixture—tending to expand and harmonize his mind, and elevate it to conceptions of the majestic, sublime, serene, and beautiful arrangements instituted by the God of nature, must possess an organization sadly deficient, or be surrounded by circumstances indeed lamentable." I would recommend the study of the honey-bee as one best calculated to awaken the interest of the indifferent. What can arrest the attention like their structure—their diligence in collecting stores for the future—their secretion of wax and moulding it into structures with a mathematical precision astonishing the profoundest philosophers—their maternal and fraternal affection in regarding the mother's every want, and assiduous care in nursing her offspring to maturity—their unaccountable display of instinct in emergencies or accidents, filling the beholder with wonder and amazement? The mind thus contemplating such astonishing operations, cannot well avoid looking beyond these results to their divine Author. Therefore let every mind that perceives one ray of light from nature's mysterious transactions, and is capable of receiving the least enjoyment therefrom, pursue the path still inviting onward in the pursuit. Every new acquisition will bring an additional satisfaction, and assist in the next attempt, which will be commenced with a renewed and constantly increasing zest; and will arise from the contemplation a wiser, better, and a nobler being, far superior to those who have never soared beyond the gratifications of the mere animal, grovelling in the dark. Is there, in the whole circle of nature's exhaustless storehouse, any one science more inviting than this? What more exalting and refining, and at the same time making a return in profits as a pecuniary reward?

What would be the result in the aggregate of all the honey produced in the flowers of the United States annually? Suppose we estimate the productions of one acre to be one pound of honey, which is but a small part of the real product in most places; yet, as a great many acres are covered with water and forest,[22] this estimate is probably enough for the average. This State (New York) contains 47,000 square miles; 640 acres in a square mile will multiply into a little more than 30,000,000, and each acre producing its pound of honey, we have the grand result of 30,000,000 lbs. of honey. If we add the States of Pennsylvania, Ohio and Michigan, we have an amount of over 126,000,000 lbs. What it might be by including all the States, those disposed may ascertain. Enough for our purpose is made clear, and that is, a small item only of an enormous amount is now secured.

Footnotes

[1] The objectors to this hypothesis will be generally found among those who are unable to give a more plausible elucidation. Those who oppose the fact that one bee is the mother of the whole family, will probably be in the same class.

[2] See Appendix of Cottage Bee-keeper, page 118.

[3] When Mr. Miner wrote his manual recommending this size, 1,728 inches, for all situations, it should be remembered he lived on Long Island. Since removing to Oneida County in this State, either his own experience or *some other cause* has changed his views, as he now recommends my size, viz., 2,000 inches.

[4] I have added a side box occasionally, but it has seldom paid me for the trouble.

[5] A line of bees-wax made with a guide-plate, or other means, is found to be of but little use.

[6] When the comb in our glass hive is new, and white, these operations can be seen more distinctly than when very old and dark.

[7] It is said that the bees will devour these eggs also.

[8] I have had several such. It made no difference whether the eggs were in the worker-cells or drone-cells, the brood was all drones. When in the worker-cells, (and the majority was there,) they required to be lengthened about one-third. In an occurrence of this kind, the colony of workers will rapidly diminish in number, until too few are left to protect the combs from the moth. It occurs most frequently in spring, but I once had a case the last of summer. The first indications are an unusual number of caps, or

covers of cells, being under and about the hive; the workers, instead of increasing, grow less in number. When you fear this state of things, make a thorough examination, blow under the hive some tobacco smoke, as directed in pruning, invert the hive, part the combs till you can see the brood; if the worker-cells contain drones, they are readily perceived, as they project beyond the usual even surface, being very irregular, here and there a few, or perhaps but one sticking out. The worker-brood, when in their own cells, form nearly an even surface; so of the drones. The only remedy that I have found is to destroy this queen, and substitute another, which can be obtained in the swarming season, or in the fall, better than at other times. To find the queen, paralyze with puff-ball, &c. For directions see fall management.

[9] The botanical names are from Wood's Class-Book.

[10] In Wood's Class-book of Botany, "Order CII.," in a plate showing the parts of this plant, it is thus described: "Fig. 11, a pair of pollen masses suspended from the glands at an angle of the antheridium," &c.

One, when reading this simple botanical description, and seeing the plate, or the Botanist with his glasses, when he minutely inspects the parts, would not suspect anything fatal to bees about it.

[11] The history of insects, as published by Harpers, gives more particulars on this interesting subject.

[12] Since the foregoing was written, I have made some further observations on this subject. In August, 1852, I noticed, on passing under some willow trees, (*Salix Vitellina,*) that leaves, grass, and stones, were covered with a wet or shining substance. On looking among the branches, I found nearly all the smallest were covered with a species of large black *aphis*, apparently engaged in sucking the juices, and occasionally discharging a minute drop of a transparent liquid. I *guessed* this might be the honey-dew. As this was early in the morning, I resolved to visit this place again, as soon as the sun got up far enough to start out the bees, and see if they collected any of it. On my return I found not only bees in hundreds, but ants, hornets, and wasps. Some

were on the branches with the *aphis*, others on the leaves and larger branches. Some of them were even on the stones and grass under the trees, collecting it.

[13] It occurred the last of July.

[14] Mr. Gillman's patent for feeding bees, is based on the principle of a chemical change. It is said that the food he gives to the bees, when poured into the cells, becomes honey of the first quality. This appears extremely mysterious; for it is well understood that when a bee has filled its sack it will go to the hive, deposit its load, and return immediately for more; and will continue its labor throughout the day, or until the supply fails; each load occupying but few minutes. The time in going from the feeder to the hive is so short that a change so important is not at all probable. The nature of bees seems to be to *collect* honey, not *make* it; hence we find, when bees are gathering from clover, they store quite a different article than when from buckwheat,--or when we feed West India honey, in quantities sufficient to have it stored *pure* in the boxes, we find that it has lost none of its bad taste in passing through the sacks of our northern bees.

It appears most probable that, if Southern honey and cheap sugar form the basis of his food, (which it is said to,) that it is flavored with something to disguise the disagreeable qualities of the compound. Should this be the secret, it would seem like a waste to feed it to bees--a portion would be given to the brood, and possibly the old bees might not always refrain from sipping a little of the tempting nectar. Why not, when the compound was ready,--instead of wasting it by this process,--put it directly in market? Or, is it necessary to have it in the combs to help psychologize the consumer into the belief that it is honey of a pure quality?

[15] Perhaps Miner's cross-bar hive would do it.

[16] All stranger queens, introduced into a stock or swarm, are secured and detained in this manner by the workers, but whether *they* dispatch them, or this is a means adopted to incite them to a deadly conflict, writers do not agree, and I shall not attempt a

decision, as I never saw the bees voluntarily release a queen thus confined. But I have seen queens, when no bees interfered, rush together in a fatal rencounter, and one of them was soon left a fallen victim of the contest. 'Tis said it *never* happens that both are killed in these battles,--perhaps not. As I never saw *quite all* of these royal combats, of course I cannot decide.

[17] I was so well pleased with my success, especially with small families, that I detailed the most important points in a communication to the Dollar Newspaper, Philadelphia, published November, 1848.

[18] As an additional proof that this method of inverting hives in the house for winter is valuable, I would say that Mr. Miner, author of the American Bee-Keeper's Manual, seems fully to appreciate it. In. the fall of 1850, I communicated to him this method; giving my reasons for preferring it to the cold method recommended in his Manual. The trial of one winter, it appears, satisfied him of its superiority, so much so that within a year from that time he published an essay recommending it; but advised confining the bees with muslin, &c.

[19] It is presumed that the inexperienced will soon learn to distinguish such bees, as die from old age or natural causes, from those affected by the cold.

[20] Vide other causes of loss, a few pages back.

[21] There are sections of country where the difference in seasons is less than in this.

[22] It should not be forgotten that forest trees are valuable, especially when there is basswood, or even maple.